ENZYME BIOCHEMISTRY AND BIOTECHNOLOGY

Edited by
Olivia Berthoumieu

ArclerPress

Enzyme Biochemistry and Biotechnology

Edited by Olivia Berthoumieu

Edition 2018

ArclerPress

Published by
Arcler Press LLC
708 3rd Avenue, 6th Floor
New York
NY 10017
United States of America
www.arclerpress.com

ISBN: 978-1-68094-525-6 (Hardback)

Printed in United States of America

Library of Congress Control Number: 2017933102

About the Editor

Olivia Berthoumieu

Olivia graduated from Oxford University, where she received a PhD in Biochemistry in 2012. Her research interests span from the use of single molecule scale methods on membrane proteins for bionanotechnological applications to the characterisation of potential inhibitors of the amyloid beta peptide aggregation – one of the hallmarks of Alzheimer's disease. A post-doctoral fellow and then research engineer at CNRS (French national centre for scientific research) for three years, she now works as a scientific and medical translator.

Table of Contents

2 PROTEIN STRUCTURE AND FUNCTIONS

3 ENZYME CLASSIFICATION

4 ENZYMES ACTIVITY

5 ENZYME KINETICS

6 ALLOSTERIC REGULATION

7 OXIDOREDUCTASE

List of Abbreviations

DHAP	Dihyroxyacetone phosphate
EC number	Enzyme Commission number
ES	Enzyme–substrate
JCBN	Joint Commission on Biochemical Nomenclature
NOE	Nuclear Overhauser Effect
PHI	Phosphohexose Isomerase Dificiency
PBGS	Porphobilinogen Synthase
PDB	Protein Data Bank
TPI	Triosephosphate Isomerase Deficiency

Preface

Enzymes are macromolecules that act as organic catalysts in most of the organism's biochemical reactions that have functions indispensable to maintenance and activity of life. It is crucial to note that reaction rates of certain chemical conversions occurring in living organisms are extremely low, and catalysis is necessary to maintain reasonable time of cell development and division. Majority of functional enzymes are proteins, like trypsin, fumarase or papain. The biological processes that occur within all living organisms are chemical reactions, and most are regulated by enzymes. Without enzymes, many of these reactions would not take place at a perceptible rate. Model enzymes are monomeric, globular proteins. Majority of studies on nature of enzymatic reactions were conducted on trypsin, chymotrypsin or amylase. As every protein, to work correctly enzymes require proper folding, and, by consequence, are susceptible to deactivation by denaturation. The key to enzyme activity is a structure called active site. Interactions between residues of polypeptide chain amino acids cause them to create a structure of defined size, shape and sequence. The difference between active site and structural domain is that the latter is able to recognize (recognition site) and process a defined molecule. Whether the interaction between the molecule and recognition site is sterical (shape-based), hydrophobic (chymotrypsin) or ionic (trypsin) it is always specific and temporal. Inhibitors often use specific covalent binding to an active site, which deactivates enzyme permanently. Enzymes catalyze all aspects of cell metabolism. This includes the digestion of food, in which large nutrient molecules (such as proteins, carbohydrates, and fats) are broken down into smaller molecules; the conservation and transformation of chemical energy; and the construction of cellular macromolecules from smaller precursors. Many inherited human diseases, such as albinism and phenylketonuria, result from a deficiency of a particular enzyme. Enzymes also have valuable industrial and medical applications. The fermenting of wine, leavening of bread, curdling of cheese, and brewing of beer have been practiced from earliest times, but not until the 19th century were these reactions understood to be the result of the catalytic activity of enzymes. Since then, enzymes have assumed an increasing importance in industrial processes that involve organic chemical reactions. The uses of enzymes in medicine include killing disease-causing microorganisms, promoting wound healing, and diagnosing certain diseases.

Content Coverage

Chapter One is anticipated to present an introduction of enzymes. Like all catalysts, enzymes increase the rate of a reaction by lowering its activation energy.

Some enzymes can make their conversion of substrate to product occur many millions of times faster.

Chapter Two explores the structure and functions of proteins. Many proteins are enzymes that catalyze biochemical reactions and are vital to metabolism. Proteins also have structural or mechanical functions, such as actin and myosin in muscle and the proteins in the cytoskeleton, which form a system of scaffolding that maintains cell shape.

Chapter Three outline the numerical classification scheme for enzymes, based on the chemical reactions they catalyze. As a system of enzyme nomenclature, every EC number is associated with a recommended name for the respective enzyme.

Chapter Four is an exploration of an enzyme activity where substrate molecules bind and undergo a chemical reaction. The active site consists of residues that form temporary bonds with the substrate (binding site) and residues that catalyze a reaction of that substrate (catalytic site).

Chapter Five presents an emphasis on enzyme kinetics, which is the study of the chemical reactions that are catalyzed by enzymes. Studying an enzyme's kinetics can reveal the catalytic mechanism of the enzyme, its role in metabolism, how its activity is controlled, and how a drug or an agonist might inhibit the enzyme.

Chapter Six focuses on the allosteric regulation of an enzyme by binding an effector molecule at a site other than the enzyme's active site. An allosteric site does not bind substrate, but instead binds another molecule that affects the enzyme's regulation. When a molecule binds an allosteric site, it alters the enzyme's shape, or conformation, which then changes how the enzyme functions.

Chapter Seven elaborates the process of aerobic digestion typically used in an activated sludge treatment plant. Waste activated sludge and primary sludge are combined, where appropriate, and passed to a thickener where the solids content is increased.

CHAPTER 1

An Introduction to Enzymes

Chapter Contents

- **Introduction**
- **Etymology and History**
- **Mechanism**
- **Cofactors**
- **Kinetics**
- **Inhibition**
- **Biological Function**
- **Industrial Applications**
- **Genetic Disorder**

Enzymes are macromolecular biological catalysts. Enzymes accelerate, or catalyze, chemical reactions. The molecules at the beginning of the process upon which enzymes may act are called substrates and the enzyme converts these into different molecules, called products. Almost all metabolic processes in the cell need enzymes in order to occur at rates fast enough to sustain life.: The set of enzymes made in a cell determines which metabolic pathways occur in that cell. The study of enzymes is called enzymology.

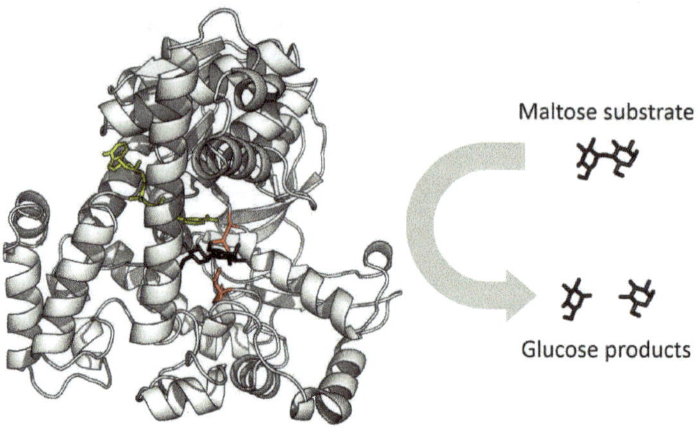

The enzyme glucosidase converts sugar maltose to two glucose sugars. Active site residues in red, maltose substrate in black, and NAD cofactor in yellow.

INTRODUCTION

Enzymes are known to catalyze more than 5,000 biochemical reaction types. Most enzymes are proteins, although a few are catalytic RNA molecules. Enzymes' specificity comes from their unique three-dimensional structures.

Like all catalysts, enzymes increase the rate of a reaction by lowering its activation energy. Some enzymes can make their conversion of substrate to product occur many millions of times faster. An extreme example is orotidine 5'-phosphate decarboxylase, which allows a reaction that would otherwise take millions of years to occur in milliseconds. Chemically, enzymes are like any catalyst and are not consumed in chemical reactions, nor do they alter the equilibrium of a reaction. Enzymes differ from most other catalysts by being much more specific. Enzyme activity can be affected by other molecules: inhibitors are molecules that decrease enzyme activity, and activators are molecules that increase activity. Many drugs and poisons are enzyme inhibitors. An enzyme's activity decreases markedly outside its optimal temperature and pH.

Some enzymes are used commercially, for example, in the synthesis of antibiotics. Some household products use enzymes to speed up chemical reactions: enzymes in biological washing powders break down protein, starch or fat stains on clothes, and enzymes in meat tenderizer break down proteins into smaller molecules, making the meat easier to chew.

ETYMOLOGY AND HISTORY

By the late 17th and early 18th centuries, the digestion of meat by stomach secretions and the conversion of starch to sugars by plant extracts and saliva were known but the mechanisms by which these occurred had not been identified.

French chemist Anselme Payen was the first to discover an enzyme, diastase, in 1833. A few decades later, when studying the fermentation of sugar to alcohol by yeast, Louis

Pasteur concluded that this fermentation was caused by a vital force contained within the yeast cells called "ferments", which were thought to function only within living organisms. He wrote that "alcoholic fermentation is an act correlated with the life and organization of the yeast cells, not with the death or putrefaction of the cells."

Eduard Buchner.

In 1877, German physiologist Wilhelm Kühne (1837–1900) first used the term enzyme, which comes from Greek ενζυμον, "leavened", to describe this process. The word enzyme was used later to refer to nonliving substances such as pepsin, and the word ferment was used to refer to chemical activity produced by living organisms.

Eduard Buchner submitted his first paper on the study of yeast extracts in 1897. In a series of experiments at the University of Berlin, he found that sugar was fermented by yeast extracts even when there were no living yeast cells in the mixture. He named the enzyme that brought about the fermentation of sucrose "zymase". In 1907, he received the Nobel Prize in Chemistry for "his discovery of cell-free fermentation". Following Buchner's example, enzymes are usually named according to the reaction they carry out: the suffix -ase is combined with the name of the substrate (e.g., lactase is the enzyme that cleaves lactose) or to the type of reaction (e.g., DNA polymerase forms DNA polymers).

The biochemical identity of enzymes was still unknown in the early 1900s. Many scientists observed that enzymatic activity was associated with proteins, but others (such as Nobel laureate Richard Willstätter) argued that proteins were merely carriers for the true enzymes and that proteins per se were incapable of catalysis.In 1926, James B. Sumner showed that the enzyme urease was a pure protein and crystallized it; he did likewise for the enzyme catalase in 1937. The conclusion that pure proteins can be enzymes was definitively demonstrated by John Howard Northrop and Wendell Meredith Stanley, who worked on the digestive enzymes pepsin (1930), trypsin and chymotrypsin. These three scientists were awarded the 1946 Nobel Prize in Chemistry.

The discovery that enzymes could be crystallized eventually allowed their structures to be solved by x-ray crystallography. This was first done for lysozyme, an enzyme found in tears, saliva and egg whites that digests the coating of some bacteria; the structure was solved by a group led by David Chilton Phillips and published in 1965. This high-resolution structure of lysozyme marked the beginning of the field of structural biology and the effort to understand how enzymes work at an atomic level of detail.

Naming Conventions

An enzyme's name is often derived from its substrate or the chemical reaction it catalyzes, with the word ending in -ase.:8.1.3 Examples are lactase, alcohol dehydrogenase and DNA polymerase. Different enzymes that catalyze the same chemical reaction are called isozymes.:10.3

The International Union of Biochemistry and Molecular Biology have developed a nomenclature for enzymes, the EC numbers; each enzyme is described by a sequence of four numbers preceded by "EC". The first number broadly classifies the enzyme based on its mechanism.

The top-level classification is:

- EC 1, Oxidoreductases: catalyze oxidation/reduction reactions
- EC 2, Transferases: transfer a functional group (*e.g.* a methyl or phosphate group)
- EC 3, Hydrolases: catalyze the hydrolysis of various bonds
- EC 4, Lyases: cleave various bonds by means other than hydrolysis and oxidation
- EC 5, Isomerases: catalyze isomerization changes within a single molecule
- EC 6, Ligases: join two molecules with covalent bonds.

These sections are subdivided by other features such as the substrate, products, and chemical mechanism. An enzyme is fully specified by four numerical designations. For example, hexokinase (EC 2.7.1.1) is a transferase (EC 2) that adds a phosphate group (EC 2.7) to a hexose sugar, a molecule containing an alcohol group (EC 2.7.1).

Structure

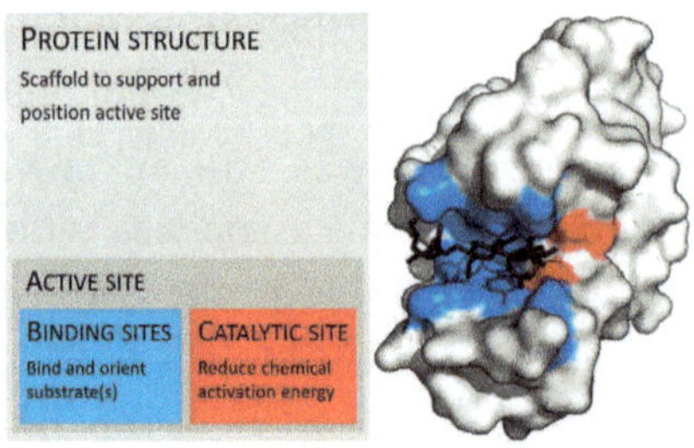

Organisation of enzyme structure and lysozyme example. Binding sites in blue, catalytic site in red and peptidoglycan substrate in black.

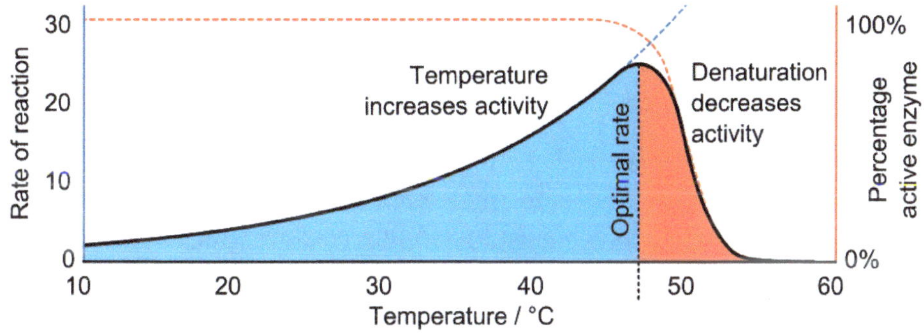

Enzyme activity initially increases with temperature (Q10 coefficient) until the enzyme's structure unfolds (denaturation), leading to an optimal rate of reaction at an intermediate temperature.

Enzymes are generally globular proteins, acting alone or in larger complexes. Like all proteins, enzymes are linear chains of amino acids that fold to produce a three-dimensional structure. The sequence of the amino acids specifies the structure which in turn determines the catalytic activity of the enzyme. Although structure determines function, a novel enzyme's activity cannot yet be predicted from its structure alone. Enzyme structures unfold (denature) when heated or exposed to chemical denaturants and this disruption to the structure typically causes a loss of activity. Enzyme denaturation is normally linked to temperatures above a species' normal level; as a result, enzymes from bacteria living in volcanic environments such as hot springs are prized by industrial users for their ability to function at high temperatures, allowing enzyme-catalysed reactions to be operated at a very high rate.

Enzymes are usually much larger than their substrates. Sizes range from just 62 amino acid residues, for the monomer of 4-oxalocrotonate tautomerase, to over 2,500 residues in the animal fatty acid synthase. Only a small portion of their structure (around 2–4 amino acids) is directly involved in catalysis: the catalytic site. This catalytic site is located next to one or more binding sites where residues orient the substrates. The catalytic site and binding site together comprise the enzyme's active site. The remaining majority of the enzyme structure serves to maintain the precise orientation and dynamics of the active site.

In some enzymes, no amino acids are directly involved in catalysis; instead, the enzyme contains sites to bind and orient catalytic cofactors. Enzyme structures may also contain allosteric sites where the binding of a small molecule causes a conformational change that increases or decreases activity.

A small number of RNA-based biological catalysts called ribozymes exist, which again can act alone or in complex with proteins. The most common of these is the ribosome which is a complex of protein and catalytic RNA components.:2.2

MECHANISM

Substrate Binding

Enzymes must bind their substrates before they can catalyse any chemical reaction. Enzymes are usually very specific as to what substrates they bind and then the chemical reaction catalysed. Specificity is achieved by binding pockets with complementary shape, charge and hydrophilic/hydrophobic characteristics to the substrates. Enzymes can therefore distinguish between very similar substrate molecules to be chemoselective, regioselective and stereospecific.

Some of the enzymes showing the highest specificity and accuracy are involved in the copying and expression of the genome. Some of these enzymes have "proof-reading" mechanisms. Here, an enzyme such as DNA polymerase catalyzes a reaction in a first step and then checks that the product is correct in a second step.This two-step process results in average error rates of less than 1 error in 100 million reactions in high-fidelity mammalian polymerases.:5.3.1 Similar proofreading mechanisms are also found in RNA polymerase, aminoacyl tRNA synthetases and ribosomes.

Conversely, some enzymes display enzyme promiscuity, having broad specificity and acting on a range of different physiologically relevant substrates. Many enzymes possess small side activities which arose fortuitously (i.e. neutrally), which may be the starting point for the evolutionary selection of a new function.

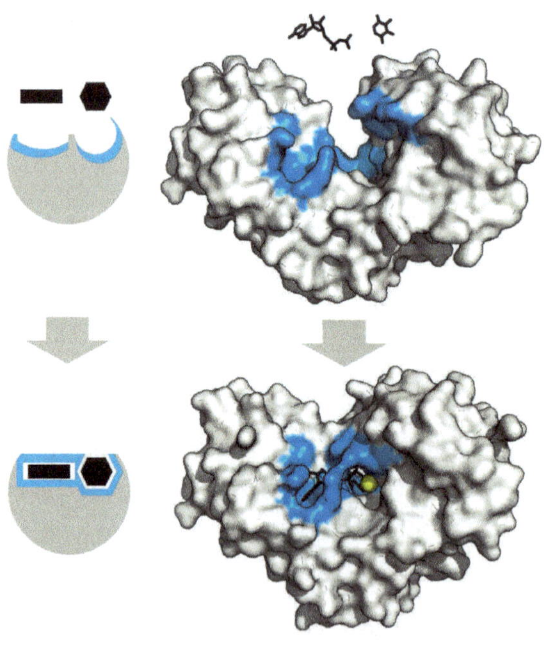

Enzyme changes shape by induced fit upon substrate binding to form enzyme-substrate complex. Hexokinase has a large induced fit motion that closes over the substrates adenosine triphosphate and xylose. Binding sites in blue, substrates in black and Mg^{2+} cofactor in yellow.

"Lock and key" model

To explain the observed specificity of enzymes, in 1894 Emil Fischer proposed that both the enzyme and the substrate possess specific complementary geometric shapes that fit exactly into one another. This is often referred to as "the lock and key" model.:8.3.2 This early model explains enzyme specificity, but fails to explain the stabilization of the transition state that enzymes achieve.

Induced fit Model

In 1958, Daniel Koshland suggested a modification to the lock and key model: since enzymes are rather flexible structures, the active site is continuously reshaped by interactions with the substrate as the substrate interacts with the enzyme. As a result, the substrate does not simply bind to a rigid active site; the amino acid side-chains that make up the active site are molded into the precise positions that enable the enzyme to perform its catalytic function. In some cases, such as glycosidases, the substrate molecule also changes shape slightly as it enters the active site. The active site continues to change until the substrate is completely bound, at which point the final shape and charge distribution is determined. Induced fit may enhance the fidelity of molecular recognition in the presence of competition and noise via the conformational proofreading mechanism.

Catalysis

Enzymes can accelerate reactions in several ways, all of which lower the activation energy ($\Delta G\ddagger$, Gibbs free energy)

- By stabilizing the transition state:
  - Creating an environment with a charge distribution complementary to that of the transition state to lower its energy.
- By providing an alternative reaction pathway:
  - Temporarily reacting with the substrate, forming a covalent intermediate to provide a lower energy transition state.
- By destabilising the substrate ground state:
  - Distorting bound substrate(s) into their transition state form to reduce the energy required to reach the transition state.
  - By orienting the substrates into a productive arrangement to reduce the reaction entropy change. The contribution of this mechanism to catalysis is relatively small.

Enzymes may use several of these mechanisms simultaneously. For example, proteases such as trypsin perform covalent catalysis using a catalytic triad, stabilise charge build-up on the transition states using an oxyanion hole, complete hydrolysis using an oriented water substrate.

Dynamics

Enzymes are not rigid, static structures; instead they have complex internal dynamic motions – that is, movements of parts of the enzyme's structure such as individual amino acid residues, groups of residues forming a protein loop or unit of secondary structure, or even an entire protein domain. These motions give rise to a conformational ensemble of slightly different structures that interconvert with one another at equilibrium. Different states within this ensemble may be associated with different aspects of an enzyme's function. For example, different conformations of the enzyme dihydrofolate reductase are associated with the substrate binding, catalysis, cofactor release, and product release steps of the catalytic cycle.

Allosteric Modulation

Allosteric sites are pockets on the enzyme, distinct from the active site, that bind to molecules in the cellular environment. These molecules then cause a change in the conformation or dynamics of the enzyme that is transduced to the active site and thus affects the reaction rate of the enzyme. In this way, allosteric interactions can either inhibit or activate enzymes. Allosteric interactions with metabolites upstream or downstream in an enzyme's metabolic pathway cause feedback regulation, altering the activity of the enzyme according to the flux through the rest of the pathway.

COFACTORS

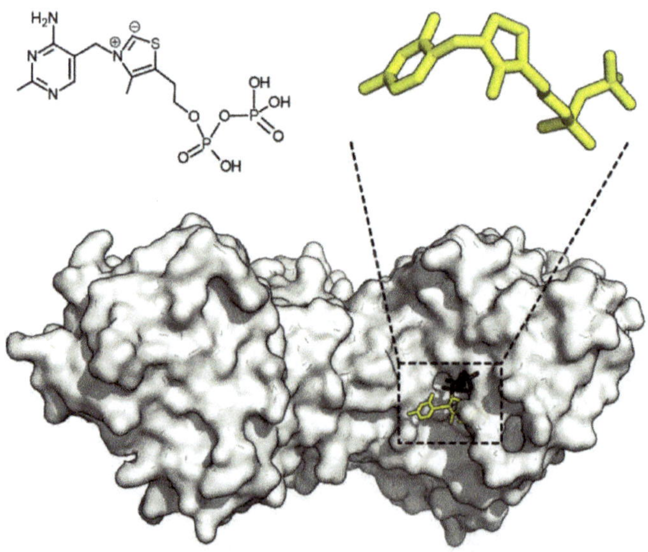

Chemical structure for thiamine pyrophosphate and protein structure of transketolase. Thiamine pyrophosphate cofactor in yellow and xylulose 5-phosphate substrate in black.

Some enzymes do not need additional components to show full activity. Others require non-protein molecules called cofactors to be bound for activity. Cofactors can be either inorganic (e.g., metal ions and iron-sulfur clusters) or organic compounds (e.g., flavin and heme). Organic cofactors can be either coenzymes, which are released from the enzyme's active site during the reaction, or prosthetic groups, which are tightly bound to an enzyme. Organic prosthetic groups can be covalently bound (e.g., biotin in enzymes such as pyruvate carboxylase).

An example of an enzyme that contains a cofactor is carbonic anhydrase, which is shown in the ribbon diagram above with a zinc cofactor bound as part of its active site. These tightly bound ions or molecules are usually found in the active site and are involved in catalysis.:8.1.1 For example, flavin and heme cofactors are often involved in redox reactions.:17

Enzymes that require a cofactor but do not have one bound are called apoenzymes or apoproteins. An enzyme together with the cofactor(s) required for activity is called a holoenzyme (or haloenzyme). The term holoenzyme can also be applied to enzymes that contain multiple protein subunits, such as the DNA polymerases; here the holoenzyme is the complete complex containing all the subunits needed for activity.:8.1.1

Coenzymes

Coenzymes are small organic molecules that can be loosely or tightly bound to an enzyme. Coenzymes transport chemical groups from one enzyme to another. Examples include NADH, NADPH and adenosine triphosphate (ATP). Some coenzymes, such as riboflavin, thiamine and folic acid, are vitamins, or compounds that cannot be synthesized by the body and must be acquired from the diet. The chemical groups carried include the hydride ion (H−) carried by NAD or NADP+, the phosphate group carried by adenosine triphosphate, the acetyl group carried by coenzyme A, formyl, methenyl or methyl groups carried by folic acid and the methyl group carried by S-adenosylmethionine.

Since coenzymes are chemically changed as a consequence of enzyme action, it is useful to consider coenzymes to be a special class of substrates, or second substrates, which are common to many different enzymes. For example, about 1000 enzymes are known to use the coenzyme NADH.

Coenzymes are usually continuously regenerated and their concentrations maintained at a steady level inside the cell. For example, NADPH is regenerated through the pentose phosphate pathway and S-adenosylmethionine by methionine adenosyltransferase. This continuous regeneration means that small amounts of coenzymes can be used very intensively. For example, the human body turns over its own weight in ATP each day.

THERMODYNAMICS

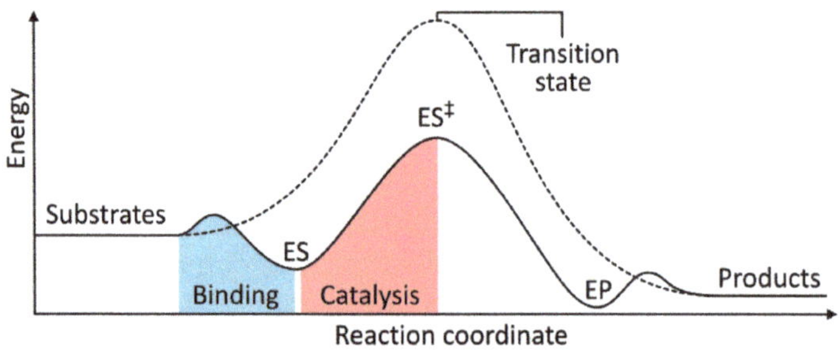

The energies of the stages of a chemical reaction. Uncatalysed (dashed line), substrates need a lot of activation energy to reach a transition state, which then decays into lower-energy products. When enzyme catalysed (solid line), the enzyme binds the substrates (ES), then stabilizes the transition state (ES‡) to reduce the activation energy required to produce products (EP) which are finally released.

As with all catalysts, enzymes do not alter the position of the chemical equilibrium of the reaction. In the presence of an enzyme, the reaction runs in the same direction as it would without the enzyme, just more quickly.:8.2.3 For example, carbonic anhydrase catalyzes its reaction in either direction depending on the concentration of its reactants:

$$CO_2 + H_2O \xrightarrow{\text{Carbonic anhydrase}} H_2CO_3$$

$$H_2CO_3 \xrightarrow{\text{Carbonic anhydrase}} CO_2 + H_2O$$

The rate of a reaction is dependent on the activation energy needed to form the transition state which then decays into products. Enzymes increase reaction rates by lowering the energy of the transition state. First, binding forms a low energy enzyme-substrate complex (ES). Secondly the enzyme stabilises the transition state such that it requires less energy to achieve compared to the uncatalyzed reaction (ES‡). Finally the enzyme-product complex (EP) dissociates to release the products.:

Enzymes can couple two or more reactions, so that a thermodynamically favorable reaction can be used to "drive" a thermodynamically unfavourable one so that the combined energy of the products is lower than the substrates. For example, the hydrolysis of ATP is often used to drive other chemical reactions.

KINETICS

Enzyme kinetics is the investigation of how enzymes bind substrates and turn them into products. The rate data used in kinetic analyses are commonly obtained from enzyme assays. In 1913 Leonor Michaelis and Maud Leonora Menten proposed a quantitative theory of enzyme kinetics, which is referred to as Michaelis–Menten kinetics. The major contribution of Michaelis and Menten was to think of enzyme reactions in two stages.

In the first, the substrate binds reversibly to the enzyme, forming the enzyme-substrate complex. This is sometimes called the Michaelis-Menten complex in their honor. The enzyme then catalyzes the chemical step in the reaction and releases the product. This work was further developed by G. E. Briggs and J. B. S. Haldane, who derived kinetic equations that are still widely used today.

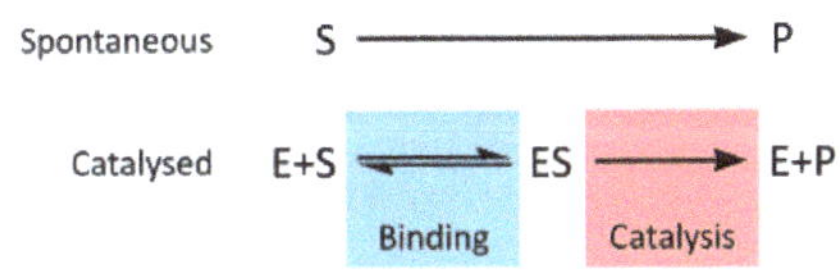

A chemical reaction mechanism with or without enzyme catalysis. The enzyme (E) binds substrate (S) to produce product (P).

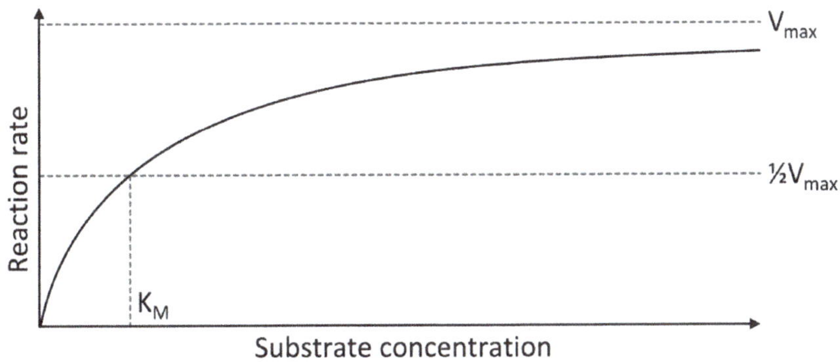

Saturation curve for an enzyme reaction showing the relation between the substrate concentration and reaction rate.

Enzyme rates depend on solution conditions and substrate concentration. To find the maximum speed of an enzymatic reaction, the substrate concentration is increased until a constant rate of product formation is seen. This is shown in the saturation curve on the right. Saturation happens because, as substrate concentration increases, more and more of the free enzyme is converted into the substrate-bound ES complex. At the maximum reaction rate (Vmax) of the enzyme, all the enzyme active sites are bound to substrate, and the amount of ES complex is the same as the total amount of enzyme.:

Vmax is only one of several important kinetic parameters. The amount of substrate needed to achieve a given rate of reaction is also important. This is given by the Michaelis-Menten constant (Km), which is the substrate concentration required for an enzyme to reach one-half its maximum reaction rate; generally, each enzyme has a characteristic Km for a given substrate. Another useful constant is kcat, also called the turnover number, which is the number of substrate molecules handled by one active site per second.:8.4

The efficiency of an enzyme can be expressed in terms of kcat/Km. This is also called the specificity constant and incorporates the rate constants for all steps in the reaction up to and including the first irreversible step. Because the specificity constant reflects both affinity and catalytic ability, it is useful for comparing different enzymes against each other, or the same enzyme with different substrates. The theoretical maximum for the specificity constant is called the diffusion limit and is about 108 to 109 (M−1 s−1). At this point every collision of the enzyme with its substrate will result in catalysis, and the rate of product formation is not limited by the reaction rate but by the diffusion rate. Enzymes with this property are called catalytically perfect or kinetically perfect. Example of such enzymes are triose-phosphate isomerase, carbonic anhydrase, acetylcholinesterase, catalase, fumarase, β-lactamase, and superoxide dismutase.:8.4.2 The turnover of such enzymes can reach several million reactions per second.:9.2

Michaelis–Menten kinetics relies on the law of mass action, which is derived from the assumptions of free diffusion and thermodynamically driven random collision. Many biochemical or cellular processes deviate significantly from these conditions, because of macromolecular crowding and constrained molecular movement.More recent, complex extensions of the model attempt to correct for these effects.

INHIBITION

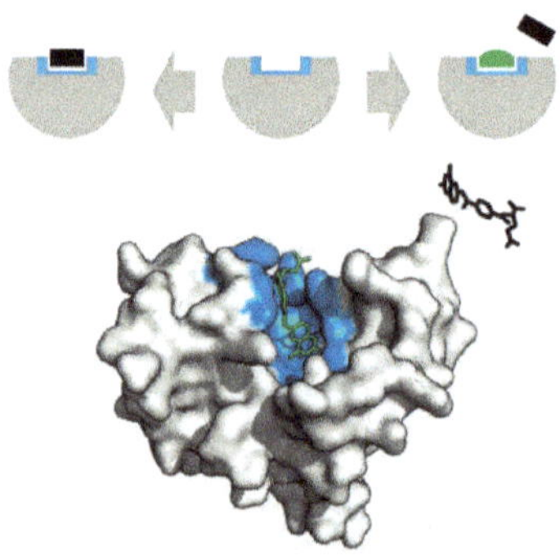

An enzyme binding site that would normally bind substrate can alternatively bind a competitive inhibitor, preventing substrate access. Dihydrofolate reductase is inhibited by methotrexate which prevents binding of its substrate, folic acid. Binding site in blue, inhibitor in green, and substrate in black.

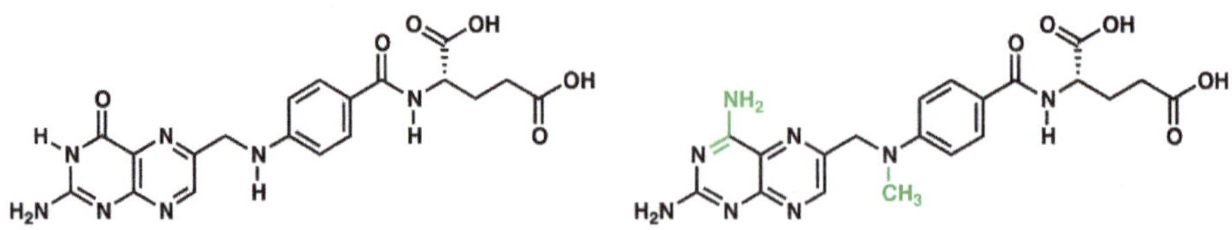

The coenzyme folic acid (left) and the anti-cancer drug methotrexate (right) are very similar in structure (differences show in green). As a result, methotrexate is a competitive inhibitor of many enzymes that use folates.

Enzyme reaction rates can be decreased by various types of enzyme inhibitors.:73–74

Types of Inhibition

Competitive

A competitive inhibitor and substrate cannot bind to the enzyme at the same time. Often competitive inhibitors strongly resemble the real substrate of the enzyme. For example, the drug methotrexate is a competitive inhibitor of the enzyme dihydrofolate reductase, which catalyzes the reduction of dihydrofolate to tetrahydrofolate. The similarity between the structures of dihydrofolate and this drug are shown in the accompanying figure. This type of inhibition can be overcome with high substrate concentration. In some cases, the inhibitor can bind to a site other than the binding-site of the usual substrate and exert an allosteric effect to change the shape of the usual binding-site.

Non-competitive

A non-competitive inhibitor binds to a site other than where the substrate binds. The substrate still binds with its usual affinity and hence Km remains the same. However the inhibitor reduces the catalytic efficiency of the enzyme so that Vmax is reduced. In contrast to competitive inhibition, non-competitive inhibition cannot be overcome with high substrate concentration.

Uncompetitive

An uncompetitive inhibitor cannot bind to the free enzyme, only to the enzyme-substrate complex; hence, these types of inhibitors are most effective at high substrate concentration. In the presence of the inhibitor, the enzyme-substrate complex is inactive.:78 This type of inhibition is rare.

Mixed

A mixed inhibitor binds to an allosteric site and the binding of the substrate and the inhibitor affect each other. The enzyme's function is reduced but not eliminated when bound to the inhibitor. This type of inhibitor does not follow the Michaelis-Menten equation.

Irreversible

An irreversible inhibitor permanently inactivates the enzyme, usually by forming a covalent bond to the protein. Penicillin and aspirin are common drugs that act in this manner.

Functions of Inhibitors

In many organisms, inhibitors may act as part of a feedback mechanism. If an enzyme produces too much of one substance in the organism, that substance may act as an inhibitor for the enzyme at the beginning of the pathway that produces it, causing production of the substance to slow down or stop when there is sufficient amount. This is a form of negative feedback. Major metabolic pathways such as the citric acid cycle make use of this mechanism.

Since inhibitors modulate the function of enzymes they are often used as drugs. Many such drugs are reversible competitive inhibitors that resemble the enzyme's native substrate, similar to methotrexate above; other well-known examples include statins used to treat high cholesterol, and protease inhibitors used to treat retroviral infections such as HIV. A common example of an irreversible inhibitor that is used as a drug is aspirin, which inhibits the COX-1 and COX-2 enzymes that produce the inflammation messenger prostaglandin. Other enzyme inhibitors are poisons. For example, the poison cyanide is an irreversible enzyme inhibitor that combines with the copper and iron in the active site of the enzyme cytochrome c oxidase and blocks cellular respiration.

BIOLOGICAL FUNCTION

Enzymes serve a wide variety of functions inside living organisms. They are indispensable for signal transduction and cell regulation, often via kinases and phosphatases. They also generate movement, with myosin hydrolyzing ATP to generate muscle contraction, and also transport cargo around the cell as part of the cytoskeleton. Other ATPases in the cell membrane are ion pumps involved in active transport. Enzymes are also involved in more exotic functions, such as luciferase generating light in fireflies. Viruses can also contain enzymes for infecting cells, such as the HIV integrase and reverse transcriptase, or for viral release from cells, like the influenza virus neuraminidase.

An important function of enzymes is in the digestive systems of animals. Enzymes such as amylases and proteases break down large molecules (starch or proteins, respectively) into smaller ones, so they can be absorbed by the intestines. Starch molecules, for example, are too large to be absorbed from the intestine, but enzymes hydrolyze the starch chains into smaller molecules such as maltose and eventually glucose, which can then be absorbed. Different enzymes digest different food substances. In ruminants, which have herbivorous diets, microorganisms in the gut produce another enzyme, cellulase, to break down the cellulose cell walls of plant fiber.

Metabolism

Several enzymes can work together in a specific order, creating metabolic pathways.:30.1 In a metabolic pathway, one enzyme takes the product of another enzyme as a substrate. After the catalytic reaction, the product is then passed on to another enzyme. Sometimes more than one enzyme can catalyze the same reaction in parallel; this can allow more complex regulation: with, for example, a low constant activity provided by one enzyme but an inducible high activity from a second enzyme.

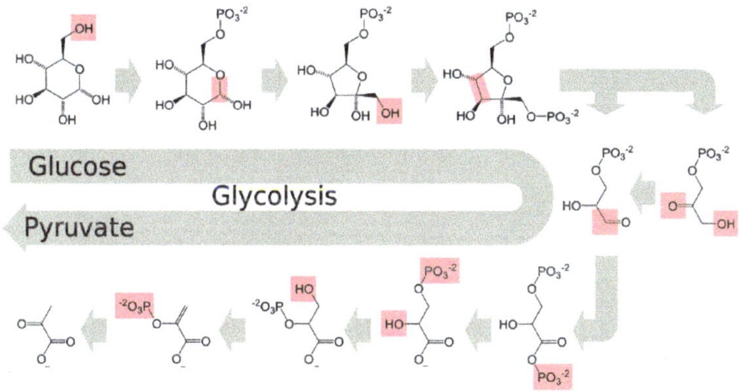

The metabolic pathway of glycolysis releases energy by converting glucose to pyruvate by via a series of intermediate metabolites. Each chemical modification (red box) is performed by a different enzyme.

Enzymes determine what steps occur in these pathways. Without enzymes, metabolism would neither progress through the same steps and could not be regulated to serve the needs of the cell. Most central metabolic pathways are regulated at a few key steps, typically through enzymes whose activity involves the hydrolysis of ATP. Because this reaction releases so much energy, other reactions that are thermodynamically unfavorable can be coupled to ATP hydrolysis, driving the overall series of linked metabolic reactions.:30.1

Control of Activity

There are five main ways that enzyme activity is controlled in the cell.

Regulation

Enzymes can be either activated or inhibited by other molecules. For example, the end product(s) of a metabolic pathway are often inhibitors for one of the first enzymes of the pathway (usually the first irreversible step, called committed step), thus regulating the amount of end product made by the pathways. Such a regulatory mechanism is called a negative feedback mechanism, because the amount of the end product produced is regulated by its own concentration.Negative feedback mechanism can effectively adjust the rate of synthesis of intermediate metabolites according to the demands of the cells. This helps with effective allocations of materials and energy economy, and it prevents the excess manufacture of end products. Like other homeostatic devices, the control of enzymatic action helps to maintain a stable internal environment in living organisms.:

Post-Translational Modification

Examples of post-translational modification include phosphorylation, myristoylation and glycosylation. For example, in the response to insulin, the phosphorylation of multiple enzymes, including glycogen synthase, helps control the synthesis or degradation of glycogen and allows the cell to respond to changes in blood sugar. Another example of post-translational modification is the cleavage of the polypeptide chain. Chymotrypsin, a digestive protease, is produced in inactive form as chymotrypsinogen in the pancreas and transported in this form to the stomach where it is activated. This stops the enzyme from digesting the pancreas or other tissues before it enters the gut. This type of inactive precursor to an enzyme is known as a zymogen or proenzyme.

Quantity

Enzyme production (transcription and translation of enzyme genes) can be enhanced or diminished by a cell in response to changes in the cell's environment. This form of gene regulation is called enzyme induction. For example, bacteria may become resistant to antibiotics such as penicillin because enzymes called beta-lactamases are induced that hydrolyse the crucial beta-lactam ring within the penicillin molecule. Another example comes from enzymes in the liver called cytochrome P450 oxidases, which are important in drug metabolism. Induction or inhibition of these enzymes can cause drug interactions. Enzyme levels can also be regulated by changing the rate of enzyme degradation.

Subcellular Distribution

Enzymes can be compartmentalized, with different metabolic pathways occurring in different cellular compartments. For example, fatty acids are synthesized by one set of enzymes in the cytosol, endoplasmic reticulum and Golgi and used by a different set of enzymes as a source of energy in the mitochondrion, through β-oxidation. In addition, trafficking of the enzyme to different compartments may change the degree of protonation (cytoplasm neutral and lysosome acidic) or oxidative state [e.g., oxidized (periplasm) or reduced (cytoplasm)] which in turn affects enzyme activity.

Organ Specialization

In multicellular eukaryotes, cells in different organs and tissues have different patterns of gene expression and therefore have different sets of enzymes (known as isozymes) available for metabolic reactions. This provides a mechanism for regulating the overall metabolism of the organism. For example, hexokinase, the first enzyme in the glycolysis pathway, has a specialized form called glucokinase expressed in the liver and pancreas that has a lower affinity for glucose yet is more sensitive to glucose concentration. This enzyme is involved in sensing blood sugar and regulating insulin production.

Involvement in Disease

Since the tight control of enzyme activity is essential for homeostasis, any malfunction (mutation, overproduction, underproduction or deletion) of a single critical enzyme can lead to a genetic disease. The malfunction of just one type of enzyme out of the thousands of types present in the human body can be fatal. An example of a fatal genetic disease due to enzyme insufficiency is Tay-Sachs disease, in which patients lack the enzyme hexosaminidase.

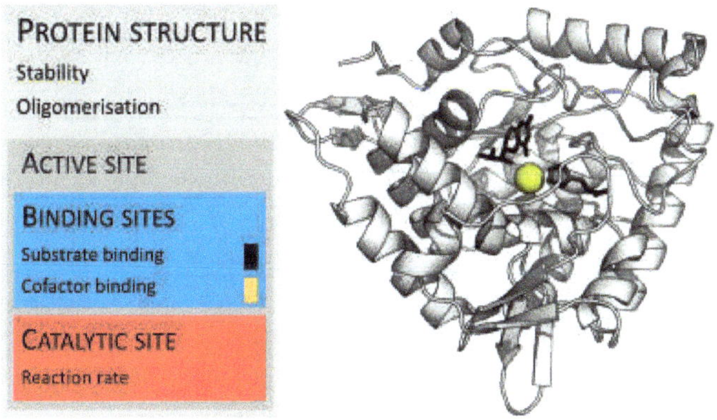

In phenylalanine hydroxylase over 300 different mutations throughout the structure cause phenylketonuria. Phenylalanine substrate and tetrahydrobiopterin coenzyme in black, and Fe^{2+} cofactor in yellow.

One example of enzyme deficiency is the most common type of phenylketonuria. Many different single amino acid mutations in the enzyme phenylalanine hydroxylase, which catalyzes the first step in the degradation of phenylalanine, result in build-up of phenylalanine and related products. Some mutations are in the active site, directly disrupting binding and catalysis, but many are far from the active site and reduce activity by destabilising the protein structure, or affecting correct oligomerisation. This can lead to intellectual disability if the disease is untreated. Another example is pseudocholinesterase deficiency, in which the body's ability to break down choline ester drugs is impaired. Oral administration of enzymes can be used to treat some functional enzyme deficiencies, such as pancreatic insufficiency and lactose intolerance.

Another way enzyme malfunctions can cause disease comes from germline mutations in genes coding for DNA repair enzymes. Defects in these enzymes cause cancer because cells are less able to repair mutations in their genomes. This causes a slow accumulation of mutations and results in the development of cancers. An example of such a hereditary cancer syndrome is xeroderma pigmentosum, which causes the development of skin cancers in response to even minimal exposure to ultraviolet light.

INDUSTRIAL APPLICATIONS

Enzymes are used in the chemical industry and other industrial applications when extremely specific catalysts are required. Enzymes in general are limited in the number of reactions they have evolved to catalyze and also by their lack of stability in organic solvents and at high temperatures. As a consequence, protein engineering is an active area of research and involves attempts to create new enzymes with novel properties, either through rational design or in vitro evolution. These efforts have begun to be successful, and a few enzymes have now been designed "from scratch" to catalyze reactions that do not occur in nature.

Application	Enzymes used	Uses
Biofuel industry	Cellulases	Break down cellulose into sugars that can be fermented to produce cellulosic ethanol.
	Ligninases	Pretreatment of biomass for biofuel production.
Biological detergent	Proteases, amylases, lipases	Remove protein, starch, and fat or oil stains from laundry and dishware.
	Mannanases	Remove food stains from the common food additive guar gum.
Brewing industry	Amylase, glucanases, proteases	Split polysaccharides and proteins in the malt.
	Betaglucanases	Improve the wort and beer filtration characteristics.
	Amyloglucosidase and pullulanases	Make low-calorie beer and adjust fermentability.
	Acetolactate decarboxylase (ALDC)	Increase fermentation efficiency by reducing diacetyl formation.
Culinary uses	Papain	Tenderize meat for cooking.
Dairy industry	Rennin	Hydrolyze protein in the manufacture of cheese.
	Lipases	Produce Camembert cheese and blue cheeses such as Roquefort.
Food processing	Amylases	Produce sugars from starch, such as in making high-fructose corn syrup.
	Proteases	Lower the protein level of flour, as in biscuit-making.
	Trypsin	Manufacture hypoallergenic baby foods.
	Cellulases, pectinases	Clarify fruit juices.
Molecular biology	Nucleases, DNA ligase and polymerases	Use restriction digestion and the polymerase chain reaction to create recombinant DNA.
Paper industry	Xylanases, hemicellulases and lignin peroxidases	Remove lignin from kraft pulp.
Personal care	Proteases	Remove proteins on contact lenses to prevent infections.
Starch industry	Amylases	Convert starch into glucose and various syrups.

GENETIC DISORDER

A genetic disorder is a genetic problem caused by one or more abnormalities in the genome, especially a condition that is present from birth (congenital). Most genetic disorders are quite rare and affect one person in every several thousands or millions.

Genetic disorders may be hereditary, passed down from the parents' genes. In other genetic disorders, defects may be caused by new mutations or changes to the DNA. In such cases, the defect will only be passed down if it occurs in the germ line. The same disease, such as some forms of cancer, may be caused by an inherited genetic condition in some people, by new mutations in other people, and mainly by environmental causes in other people. Whether, when and to what extent a person with the genetic defect or abnormality will actually suffer from the disease is almost always affected by the environmental factors and events in the person's development.

Some types of recessive gene disorders confer an advantage in certain environments when only one copy of the gene is present.

Single-Gene

A single-gene disorder is the result of a single mutated gene. Over 4000 human diseases are caused by single-gene defects. Single-gene disorders can be passed on to subsequent generations in several ways. Genomic imprinting and uniparental disomy, however, may affect inheritance patterns.

Disorder prevalence (approximate)	
Autosomal dominant	
Familial hypercholesterolemia	1 in 500
Polycystic kidney disease	1 in 1250
Neurofibromatosis type I	1 in 2,500
Hereditary spherocytosis	1 in 5,000
Marfan syndrome	1 in 4,000[2]
Huntington's disease	1 in 15,000[3]
Autosomal recessive	
Sickle cell anaemia	1 in 625
Cystic fibrosis	1 in 2,000
Tay-Sachs disease	1 in 3,000
Phenylketonuria	1 in 12,000
Mucopolysaccharidoses	1 in 25,000
Lysosomal acid lipase deficiency	1 in 40,000
Glycogen storage diseases	1 in 50,000
Galactosemia	1 in 57,000
X-linked	
Duchenne muscular dystrophy	1 in 7,000
Hemophilia	1 in 10,000
Values are for liveborn infants	

The divisions between recessive and dominant types are not "hard and fast", although the divisions between autosomal and X-linked types are (since the latter types are distinguished purely based on the chromosomal location of the gene). For example, achondroplasia is typically considered a dominant disorder, but children with two genes for achondroplasia have a severe skeletal disorder of which achondroplasics could be viewed as carriers. Sickle-cell anemia is also considered a recessive condition, but heterozygous carriers have increased resistance to malaria in early childhood, which could be described as a related dominant condition.When a couple where one partner or both are sufferers or carriers of a single-gene disorder wish to have a child, they can do so through in vitro fertilization, which means they can then have a preimplantation genetic diagnosis to check whether the embryo has the genetic disorder.

Autosomal dominant

Only one mutated copy of the gene will be necessary for a person to be affected by an autosomal dominant disorder. Each affected person usually has one affected parent. The chance a child will inherit the mutated gene is 50%. Autosomal dominant conditions sometimes have reduced penetrance, which means although only one mutated copy is needed, not all individuals who inherit that mutation go on to develop the disease. Examples of this type of disorder are Huntington's disease, neurofibromatosis type 1, neurofibromatosis type 2, Marfan syndrome, hereditary nonpolyposis colorectal cancer, hereditary multiple exostoses (a highly penetrant autosomal dominant disorder), Tuberous sclerosis, Von Willebrand disease, and acute intermittent porphyria. Birth defects are also called congenital anomalies.

Autosomal Recessive

Two copies of the gene must be mutated for a person to be affected by an autosomal recessive disorder. An affected person usually has unaffected parents who each carry a single copy of the mutated gene (and are referred to as carriers). Two unaffected people who each carry one copy of the mutated gene have a 25% risk with each pregnancy of having a child affected by the disorder. Examples of this type of disorder are Acrodermatitis Enteropathica, Albinism, Medium-chain acyl-CoA dehydrogenase deficiency, cystic fibrosis, sickle-cell disease, Tay-Sachs disease, Niemann-Pick disease, spinal muscular atrophy, and Roberts syndrome. Certain other phenotypes, such as wet versus dry earwax, are also determined in an autosomal recessive fashion.

X-linked Dominant

X-linked dominant disorders are caused by mutations in genes on the X chromosome. Only a few disorders have this inheritance pattern, with a prime example being X-linked hypophosphatemic rickets. Males and females are both affected in these disorders, with males typically being more severely affected than females. Some X-linked dominant conditions, such as Rett syndrome, incontinentia pigmenti type 2, and Aicardi syndrome, are usually fatal in males either in utero or shortly after birth, and are therefore predominantly seen in females. Exceptions to this finding are extremely rare cases in which boys with Klinefelter syndrome (47,XXY) also inherit an X-linked dominant condition and exhibit symptoms more similar to those of a female in terms of disease severity. The chance of passing on an X-linked dominant disorder differs between men and women. The sons of a man with an X-linked dominant disorder will all be unaffected (since they receive their father's Y chromosome), and his daughters will all inherit the condition. A woman with an X-linked dominant disorder has a 50% chance of having an affected fetus with each pregnancy, although it should be noted that in cases such as incontinentia pigmenti, only female offspring are generally viable. In addition, although these conditions do not alter fertility per se, individuals with Rett syndrome or Aicardi syndrome rarely reproduce.

X-linked Recessive

X-linked recessive conditions are also caused by mutations in genes on the X chromosome. Males are more frequently affected than females, and the chance of passing on the disorder differs between men and women. The sons of a man with an X-linked recessive disorder will not be affected, and his daughters will carry one copy of the mutated gene. A woman who is a carrier of an X-linked recessive disorder (XRXr) has a 50% chance of having sons who are affected and a 50% chance of having daughters who carry one copy of the mutated gene and are therefore carriers. X-linked recessive conditions include the serious diseases hemophilia A, Duchenne muscular dystrophy, and Lesch-Nyhan syndrome, as well as common and less serious conditions such as male pattern baldness and red-green color blindness. X-linked recessive conditions can sometimes manifest in females due to skewed X-inactivation or monosomy X (Turner syndrome).

Y-linked

Y-linked disorders, also called holandric disorders, are caused by mutations on the Y chromosome. These conditions display may only be transmitted from the heterogametic sex (e.g. male humans) to offspring of the same sex. More simply, this means that Y-linked disorders in humans can only be passed from men to their sons; females can never be affected because they do not possess Y-allosomes.

Y-linked disorders are exceedingly rare but the most well-known examples typically cause infertility. Reproduction in such conditions is only possible through the circumvention of infertility by medical intervention.

Mitochondrial

This type of inheritance, also known as maternal inheritance, applies to genes in mitochondrial DNA. Because only egg cells contribute mitochondria to the developing embryo, only mothers can pass on mitochondrial conditions to their children. An example of this type of disorder is Leber's hereditary optic neuropathy.

Many Genes

Genetic disorders may also be complex, multifactorial, or polygenic, meaning they are likely associated with the effects of multiple genes in combination with lifestyles and environmental factors. Multifactorial disorders include heart disease and diabetes. Although complex disorders often cluster in families, they do not have a clear-cut pattern of inheritance. This makes it difficult to determine a person's risk of inheriting or passing on these disorders. Complex disorders are also difficult to study and treat, because the specific factors that cause most of these disorders have not yet been identified. Studies which aim to identify the cause of complex disorders can use several methodological approaches to determine genotype-phenotype associations. One method, the genotype-first approach, starts by identifying genetic variants within patients and then determining the associated clinical manifestations. This is opposed to the more traditional phenotype-first approach, and may identify causal factors that have previously been obscured by clinical heterogeneity, penetrance, and expressivity.

On a pedigree, polygenic diseases do tend to "run in families", but the inheritance does not fit simple patterns as with Mendelian diseases. But this does not mean that the genes cannot eventually be located and studied. There is also a strong environmental component to many of them (e.g., blood pressure).

Diagnosis

Due to the wide range of genetic disorders that are presently known, diagnosis of a genetic disorder is widely varied and dependent of the disorder. Most genetic disorders are diagnosed at birth or during early childhood, however some, such as Huntington's disease, can escape detection until the patient is well into adulthood.

The basic aspects of a genetic disorder rests on the inheritance of genetic material. With an in depth family history, it is possible to anticipate possible disorders in children which direct medical professionals to specific tests depending on the disorder and allow parents the chance to prepare for potential lifestyle changes, anticipate the possibility of stillbirth, or contemplate termination. Prenatal diagnosis can detect the presence of characteristic abnormalities in fetal development through ultrasound, or detect the presence of characteristic substances via invasive procedures which involve inserting probes or needles into the uterus such as in amniocentesis.

Prognosis

Not all genetic disorders directly result in death, however there are no known cures for genetic disorders. Many genetic disorders affect stages of development such as Down's Syndrome. While others result in purely physical symptoms such as Muscular Dystrophy. Other disorders, such as Huntington's Disease show no signs until adulthood. During the active time of a genetic disorder, patients mostly rely on maintaining or slowing the degradation of quality of life and maintain patient autonomy. This includes physical therapy, pain management, and may include a selection of alternative medicine programs.

Treatment

The treatment of genetic disorders is an ongoing battle with over 1800 gene therapy clinical trials having been completed, are ongoing, or have been approved worldwide. Despite this, most treatment options revolve around treating the symptoms of the disorders in an attempt to improve patient quality of life.

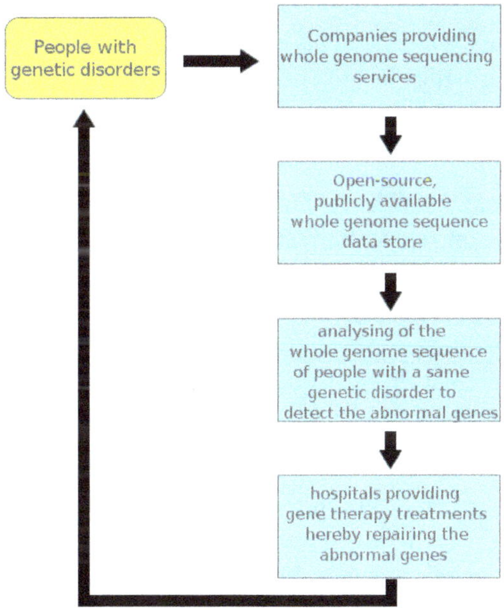

From personal genomics to gene therapy

Gene therapy refers to a form of treatment where a healthy gene is introduced to a patient. This should alleviate the defect caused by a faulty gene or slow the progression of disease. A major obstacle has been the delivery of genes to the appropriate cell, tissue, and organ affected by the disorder. How does one introduce a gene into the potentially trillions of cells which carry the defective copy? This question has been the roadblock between understanding the genetic disorder and correcting the genetic disorder.

REFERENCES

1. Stryer L, Berg JM, Tymoczko JL (2002). Biochemistry (5th ed.). San Francisco: W.H. Freeman. ISBN 0-7167-4955-6.
2. Schomburg I, Chang A, Placzek S, Söhngen C, Rother M, Lang M, Munaretto C, Ulas S, Stelzer M, Grote A, Scheer M, Schomburg D (January 2013). "BRENDA in 2013: integrated reactions, kinetic data, enzyme function data, improved disease classification: new options and contents in BRENDA". Nucleic Acids Research. 41 (Database issue): D764–72. doi:10.1093/nar/gks1049. PMC 3531171. PMID 23203881.
3. Radzicka A, Wolfenden R (January 1995). "A proficient enzyme". Science. 267 (5194): 90–931. Bibcode:1995Sci...267...90R. doi:10.1126/science.7809611. PMID 7809611.
4. Callahan BP, Miller BG (December 2007). "OMP decarboxylase—An enigma persists". Bioorganic Chemistry. 35 (6): 465–9. doi:10.1016/j.bioorg.2007.07.004.

PMID 17889251.

5. de Réaumur RA (1752). "Observations sur la digestion des oiseaux". Histoire de l'academie royale des sciences. 1752: 266, 461.
6. Williams HS (1904). A History of Science: in Five Volumes. Volume IV: Modern Development of the Chemical and Biological Sciences. Harper and Brothers.
7. Payen A, Persoz JF (1833). "Mémoire sur la diastase, les principaux produits de ses réactions et leurs applications aux arts industriels" [Memoir on diastase, the principal products of its reactions, and their applications to the industrial arts]. Annales de chimie et de physique. 2nd (in French). 53: 73–92.
8. Manchester KL (December 1995). "Louis Pasteur (1822–1895)–chance and the prepared mind". Trends in Biotechnology. 13 (12): 511–5. doi:10.1016/S0167-7799(00)89014-9. PMID 8595136.
9. Kühne coined the word "enzyme" in: Kühne W (1877). "Über das Verhalten verschiedener organisirter und sog. ungeformter Fermente" [On the behavior of various organized and so-called unformed ferments]. Verhandlungen des naturhistorisch-medicinischen Vereins zu Heidelberg. new series (in German). 1 (3): 190–193. The relevant passage occurs on page 190: «Um Missverständnissen vorzubeugen und lästige Umschreibungen zu vermeiden schlägt Vortragender vor, die ungeformten oder nicht organisirten Fermente, deren Wirkung ohne Anwesenheit von Organismen und ausserhalb derselben erfolgen kann, als Enzyme zu bezeichnen." (Translation: In order to obviate misunderstandings and avoid cumbersome periphrases, [the author, a university lecturer] suggests designating as "enzymes" the unformed or not organized ferments, whose action can occur without the presence of organisms and outside of the same.)
10. Holmes FL (2003). "Enzymes". In Heilbron JL. The Oxford Companion to the History of Modern Science. Oxford: Oxford University Press. p. 270.
11. Eduard Buchner". Nobel Laureate Biography. Nobelprize.org. Retrieved 23 February 2015.
12. Eduard Buchner – Nobel Lecture: Cell-Free Fermentation". Nobelprize.org. 1907. Retrieved 23 February 2015.
13. The naming of enzymes by adding the suffix "-ase" to the substrate on which the enzyme acts, has been traced to French scientist Émile Duclaux (1840–1904), who intended to honor the discoverers of diastase – the first enzyme to be isolated – by introducing this practice in his book Duclaux E (1899). Traité de microbiologie: Diastases, toxines et venins [Microbiology Treatise: diastases , toxins and venoms] (in French). Paris, France: Masson and Co. See Chapter 1, especially page 9.
14. Willstätter R (1927). "Faraday lecture. Problems and methods in enzyme research". Journal of the Chemical Society (Resumed): 1359. doi:10.1039/JR9270001359. quoted in Blow D (April 2000). "So do we understand how enzymes work?" (pdf). Structure (London, England : 1993). 8 (4): R77–R81. doi:10.1016/S0969-2126(00)00125-8. PMID 10801479.

15. «Nobel Prizes and Laureates: The Nobel Prize in Chemistry 1946". Nobelprize.org. Retrieved 23 February 2015.

16. Blake CC, Koenig DF, Mair GA, North AC, Phillips DC, Sarma VR (May 1965). "Structure of hen egg-white lysozyme. A three-dimensional Fourier synthesis at 2 Ångström resolution". Nature. 206 (4986): 757–61. Bibcode:1965Natur.206..757B. doi:10.1038/206757a0. PMID 5891407.

17. Johnson LN, Petsko GA (1999). "David Phillips and the origin of structural enzymology". Trends Biochem. Sci. 24 (7): 287–9. doi:10.1016/S0968-0004(99)01423-1. PMID 10390620.

18. Nomenclature Committee. "Classification and Nomenclature of Enzymes by the Reactions they Catalyse". International Union of Biochemistry and Molecular Biology (NC-IUBMB). School of Biological and Chemical Sciences, Queen Mary, University of London.

19. Nomenclature Committee. "EC 2.7.1.1". International Union of Biochemistry and Molecular Biology (NC-IUBMB). School of Biological and Chemical Sciences, Queen Mary, University of London.

20. Anfinsen CB (July 1973). "Principles that govern the folding of protein chains". Science. 181 (4096): 223–30. Bibcode:1973Sci...181..223A. doi:10.1126/science.181.4096.223. PMID 4124164.

21. Dunaway-Mariano D (November 2008). "Enzyme function discovery". Structure (London, England : 1993). 16 (11): 1599–600. doi:10.1016/j.str.2008.10.001. PMID 19000810.

22. Petsko GA, Ringe D (2003). "Chapter 1: From sequence to structure". Protein structure and function. London: New Science. p. 27. ISBN 978-1405119221.

23. Chen LH, Kenyon GL, Curtin F, Harayama S, Bembenek ME, Hajipour G, Whitman CP (September 1992). "4-Oxalocrotonate tautomerase, an enzyme composed of 62 amino acid residues per monomer". The Journal of Biological Chemistry. 267 (25): 17716–21. PMID 1339435.

24. Smith S (December 1994). "The animal fatty acid synthase: one gene, one polypeptide, seven enzymes". FASEB Journal. 8 (15): 1248–59. PMID 8001737.

25. «The Catalytic Site Atlas". The European Bioinformatics Institute. Retrieved 4 April 2007.

26. Suzuki H (2015). "Chapter 7: Active Site Structure". How Enzymes Work: From Structure to Function. Boca Raton, FL: CRC Press. pp. 117–140. ISBN 978-981-4463-92-8.

27. Krauss G (2003). "The Regulations of Enzyme Activity". Biochemistry of Signal Transduction and Regulation (3rd ed.). Weinheim: Wiley-VCH. pp. 89–114. ISBN 9783527605767.

28. Jaeger KE, Eggert T (August 2004). "Enantioselective biocatalysis optimized by directed evolution". Current Opinion in Biotechnology. 15 (4): 305–13. doi:10.1016/j.copbio.2004.06.007. PMID 15358000.

29. Shevelev IV, Hübscher U (May 2002). "The 3' 5' exonucleases". Nature Reviews

Molecular Cell Biology. 3 (5): 364–76. doi:10.1038/nrm804. PMID 11988770.

30. Zenkin N, Yuzenkova Y, Severinov K (July 2006). "Transcript-assisted transcriptional proofreading". Science. 313 (5786): 518–20. Bibcode:2006Sci...313..518Z. doi:10.1126/science.1127422. PMID 16873663.

31. Ibba M, Soll D (2000). "Aminoacyl-tRNA synthesis". Annual Review of Biochemistry. 69: 617–50. doi:10.1146/annurev.biochem.69.1.617. PMID 10966471.

32. Rodnina MV, Wintermeyer W (2001). "Fidelity of aminoacyl-tRNA selection on the ribosome: kinetic and structural mechanisms". Annual Review of Biochemistry. 70: 415–35. doi:10.1146/annurev.biochem.70.1.415. PMID 11395413.

33. Khersonsky O, Tawfik DS (2010). "Enzyme promiscuity: a mechanistic and evolutionary perspective". Annual Review of Biochemistry. 79: 471–505. doi:10.1146/annurev-biochem-030409-143718. PMID 20235827.

34. O'Brien PJ, Herschlag D (April 1999). "Catalytic promiscuity and the evolution of new enzymatic activities". Chemistry & Biology. 6 (4): R91–R105. doi:10.1016/S1074-5521(99)80033-7. PMID 10099128.

35. Fischer E (1894). "Einfluss der Configuration auf die Wirkung der Enzyme" [Influence of configuration on the action of enzymes]. Berichte der Deutschen chemischen Gesellschaft zu Berlin (in German). 27 (3): 2985–93. doi:10.1002/cber.18940270364. From page 2992: «Um ein Bild zu gebrauchen, will ich sagen, dass Enzym und Glucosid wie Schloss und Schlüssel zu einander passen müssen, um eine chemische Wirkung auf einander ausüben zu können." (To use an image, I will say that an enzyme and a glucoside [i.e., glucose derivative] must fit like a lock and key, in order to be able to exert a chemical effect on each other.)

36. Cooper GM (2000). "Chapter 2.2: The Central Role of Enzymes as Biological Catalysts". The Cell: a Molecular Approach (2nd ed.). Washington (DC): ASM Press. ISBN 0-87893-106-6.

37. Koshland DE (February 1958). "Application of a Theory of Enzyme Specificity to Protein Synthesis". Proceedings of the National Academy of Sciences of the United States of America. 44 (2): 98–104. Bibcode:1958PNAS...44...98K. doi:10.1073/pnas.44.2.98. PMC 335371. PMID 16590179.

38. Vasella A, Davies GJ, Böhm M (October 2002). "Glycosidase mechanisms". Current Opinion in Chemical Biology. 6 (5): 619–29. doi:10.1016/S1367-5931(02)00380-0. PMID 12413546.

39. Boyer R (2002). "Chapter 6: Enzymes I, Reactions, Kinetics, and Inhibition". Concepts in Biochemistry (2nd ed.). New York, Chichester, Weinheim, Brisbane, Singapore, Toronto.: John Wiley & Sons, Inc. pp. 137–8. ISBN 0-470-00379-0. OCLC 51720783.

40. Savir Y, Tlusty T (2007). Scalas E, ed. "Conformational proofreading: the impact of conformational changes on the specificity of molecular recognition" (PDF). PLoS ONE. 2 (5): e468. Bibcode:2007PLoSO...2..468S. doi:10.1371/journal.pone.0000468. PMC 1868595. PMID 17520027.

41. Fersht A (1985). Enzyme Structure and Mechanism. San Francisco: W.H. Freeman. pp. 50–2. ISBN 0-7167-1615-1.

42. Warshel A, Sharma PK, Kato M, Xiang Y, Liu H, Olsson MH (August 2006). "Electrostatic basis for enzyme catalysis". Chemical Reviews. 106 (8): 3210–35. doi:10.1021/cr0503106. PMID 16895325.

43. Cox MM, Nelson DL (2013). "Chapter 6.2: How enzymes work". Lehninger Principles of Biochemistry (6th ed.). New York, N.Y.: W.H. Freeman. p. 195. ISBN 978-1464109621.

44. Benkovic SJ, Hammes-Schiffer S (August 2003). "A perspective on enzyme catalysis". Science. 301 (5637): 1196–202. Bibcode:2003Sci...301.1196B. doi:10.1126/science.1085515. PMID 12947189.

45. Jencks WP (1987). Catalysis in Chemistry and Enzymology. Mineola, N.Y: Dover. ISBN 0-486-65460-5.

46. Villa J, Strajbl M, Glennon TM, Sham YY, Chu ZT, Warshel A (October 2000). "How important are entropic contributions to enzyme catalysis?". Proceedings of the National Academy of Sciences of the United States of America. 97 (22): 11899–904. Bibcode:2000PNAS...9711899V. doi:10.1073/pnas.97.22.11899. PMC 17266. PMID 11050223.

47. Ramanathan A, Savol A, Burger V, Chennubhotla CS, Agarwal PK (2014). "Protein conformational populations and functionally relevant substates". Acc. Chem. Res. 47 (1): 149–56. doi:10.1021/ar400084s. PMID 23988159.

48. Tsai CJ, Del Sol A, Nussinov R (2009). "Protein allostery, signal transmission and dynamics: a classification scheme of allosteric mechanisms". Mol Biosyst. 5 (3): 207–16. doi:10.1039/b819720b. PMC 2898650. PMID 19225609.

49. Changeux JP, Edelstein SJ (June 2005). "Allosteric mechanisms of signal transduction". Science. 308 (5727): 1424–8. Bibcode:2005Sci...308.1424C. doi:10.1126/science.1108595. PMID 15933191.

50. de Bolster M (1997). "Glossary of Terms Used in Bioinorganic Chemistry: Cofactor". International Union of Pure and Applied Chemistry. Retrieved 30 October 2007.

51. Chapman-Smith A, Cronan JE (1999). "The enzymatic biotinylation of proteins: a post-translational modification of exceptional specificity". Trends Biochem. Sci. 24 (9): 359–63. doi:10.1016/s0968-0004(99)01438-3. PMID 10470036.

52. Fisher Z, Hernandez Prada JA, Tu C, Duda D, Yoshioka C, An H, Govindasamy L, Silverman DN, McKenna R (February 2005). "Structural and kinetic characterization of active-site histidine as a proton shuttle in catalysis by human carbonic anhydrase II". Biochemistry. 44 (4): 1097–115. doi:10.1021/bi0480279. PMID 15667203.

53. Wagner AL (1975). Vitamins and Coenzymes. Krieger Pub Co. ISBN 0-88275-258-8.

54. "BRENDA The Comprehensive Enzyme Information System". Technische Universität Braunschweig. Retrieved 23 February 2015.

55. Törnroth-Horsefield S, Neutze R (December 2008). "Opening and closing the metabolite gate". Proceedings of the National Academy of Sciences of the United States of America. 105 (50): 19565–6. Bibcode:2008PNAS..10519565T. doi:10.1073/pnas.0810654106. PMC 2604989. PMID 19073922.

56. McArdle WD, Katch F, Katch VL (2006). "Chapter 9: The Pulmonary System and Exercise". Essentials of Exercise Physiology (3rd ed.). Baltimore, Maryland: Lippincott Williams & Wilkins. pp. 312–3. ISBN 978-0781749916.

57. Ferguson SJ, Nicholls D, Ferguson S (2002). Bioenergetics 3 (3rd ed.). San Diego: Academic. ISBN 0-12-518121-3.

58. Michaelis L, Menten M (1913). "Die Kinetik der Invertinwirkung" [The Kinetics of Invertase Action]. Biochem. Z. (in German). 49: 333–369.; Michaelis L, Menten ML, Johnson KA, Goody RS (2011). "The original Michaelis constant: translation of the 1913 Michaelis-Menten paper". Biochemistry. 50 (39): 8264–9. doi:10.1021/bi201284u. PMC 3381512. PMID 21888353.

59. Briggs GE, Haldane JB (1925). "A Note on the Kinetics of Enzyme Action". The Biochemical Journal. 19 (2): 339–339. doi:10.1042/bj0190338. PMC 1259181. PMID 16743508.

60. Ellis RJ (October 2001). "Macromolecular crowding: obvious but underappreciated". Trends in Biochemical Sciences. 26 (10): 597–604. doi:10.1016/S0968-0004(01)01938-7. PMID 11590012.

61. Kopelman R (September 1988). "Fractal reaction kinetics". Science. 241 (4873): 1620–26. Bibcode:1988Sci...241.1620K. doi:10.1126/science.241.4873.1620. PMID 17820893.

62. Cornish-Bowden A (2004). Fundamentals of Enzyme Kinetics (3 ed.). London: Portland Press. ISBN 1-85578-158-1.

63. Price NC (1979). "What is meant by 'competitive inhibition'?". Trends in Biochemical Sciences. 4 (11): N272–N273. doi:10.1016/0968-0004(79)90205-6.

64. Cornish-Bowden A (July 1986). "Why is uncompetitive inhibition so rare? A possible explanation, with implications for the design of drugs and pesticides". FEBS Letters. 203 (1): 3–6. doi:10.1016/0014-5793(86)81424-7. PMID 3720956.

65. Fisher JF, Meroueh SO, Mobashery S (February 2005). "Bacterial resistance to beta-lactam antibiotics: compelling opportunism, compelling opportunity". Chemical Reviews. 105 (2): 395–424. doi:10.1021/cr030102i. PMID 15700950.

66. Johnson DS, Weerapana E, Cravatt BF (June 2010). "Strategies for discovering and derisking covalent, irreversible enzyme inhibitors". Future Medicinal Chemistry. 2 (6): 949–64. doi:10.4155/fmc.10.21. PMC 2904065. PMID 20640225.

67. Endo A (1 November 1992). "The discovery and development of HMG-CoA reductase inhibitors" (PDF). J. Lipid Res. 33 (11): 1569–82. PMID 1464741.

68. Wlodawer A, Vondrasek J (1998). "Inhibitors of HIV-1 protease: a major success of structure-assisted drug design". Annual Review of Biophysics and Biomolecular Structure. 27: 249–84. doi:10.1146/annurev.biophys.27.1.249. PMID 9646869.

69. Yoshikawa S, Caughey WS (May 1990). "Infrared evidence of cyanide binding to iron and copper sites in bovine heart cytochrome c oxidase. Implications regarding oxygen reduction". The Journal of Biological Chemistry. 265 (14): 7945–58. PMID 2159465.

70. Hunter T (January 1995). "Protein kinases and phosphatases: the yin and yang of protein phosphorylation and signaling". Cell. 80 (2): 225–36. doi:10.1016/0092-8674(95)90405-0. PMID 7834742.

71. Berg JS, Powell BC, Cheney RE (April 2001). "A millennial myosin census". Molecular Biology of the Cell. 12 (4): 780–94. doi:10.1091/mbc.12.4.780. PMC 32266. PMID 11294886.

72. Meighen EA (March 1991). "Molecular biology of bacterial bioluminescence". Microbiological Reviews. 55 (1): 123–42. PMC 372803. PMID 2030669.

73. De Clercq E (2002). "Highlights in the development of new antiviral agents". Mini Rev Med Chem. 2 (2): 163–75. doi:10.2174/1389557024605474. PMID 12370077.

74. Mackie RI, White BA (October 1990). "Recent advances in rumen microbial ecology and metabolism: potential impact on nutrient output". Journal of Dairy Science. 73 (10): 2971–95. doi:10.3168/jds.S0022-0302(90)78986-2. PMID 2178174.

75. Rouzer CA, Marnett LJ (2009). "Cyclooxygenases: structural and functional insights". J. Lipid Res. 50 Suppl: S29–34. doi:10.1194/jlr.R800042-JLR200. PMC 2674713. PMID 18952571.

76. Suzuki H (2015). "Chapter 8: Control of Enzyme Activity". How Enzymes Work: From Structure to Function. Boca Raton, FL: CRC Press. pp. 141–69. ISBN 978-981-4463-92-8.

77. Doble BW, Woodgett JR (April 2003). "GSK-3: tricks of the trade for a multi-tasking kinase". Journal of Cell Science. 116 (Pt 7): 1175–86. doi:10.1242/jcs.00384. PMC 3006448. PMID 12615961.

78. Bennett PM, Chopra I (1993). "Molecular basis of beta-lactamase induction in bacteria" (PDF). Antimicrob. Agents Chemother. 37 (2): 153–8. doi:10.1128/aac.37.2.153. PMC 187630. PMID 8452343.

79. Skett P, Gibson GG (2001). "Chapter 3: Induction and Inhibition of Drug Metabolism". Introduction to Drug Metabolism (3 ed.). Cheltenham, UK: Nelson Thornes Publishers. pp. 87–118. ISBN 978-0748760114.

80. Faergeman NJ, Knudsen J (April 1997). "Role of long-chain fatty acyl-CoA esters in the regulation of metabolism and in cell signalling". The Biochemical Journal. 323 (Pt 1): 1–12. PMC 1218279. PMID 9173866.

81. Suzuki H (2015). "Chapter 4: Effect of pH, Temperature, and High Pressure on Enzymatic Activity". How Enzymes Work: From Structure to Function. Boca Raton, FL: CRC Press. pp. 53–74. ISBN 978-981-4463-92-8.

82. Kamata K, Mitsuya M, Nishimura T, Eiki J, Nagata Y (March 2004). "Structural basis for allosteric regulation of the monomeric allosteric enzyme human

glucokinase". Structure. 12 (3): 429–38. doi:10.1016/j.str.2004.02.005. PMID 15016359.

83. Froguel P, Zouali H, Vionnet N, Velho G, Vaxillaire M, Sun F, Lesage S, Stoffel M, Takeda J, Passa P (March 1993). "Familial hyperglycemia due to mutations in glucokinase. Definition of a subtype of diabetes mellitus". The New England Journal of Medicine. 328 (10): 697–702. doi:10.1056/NEJM199303113281005. PMID 8433729.
84. Okada S, O'Brien JS (August 1969). "Tay-Sachs disease: generalized absence of a beta-D-N-acetylhexosaminidase component". Science. 165 (3894): 698–700. Bibcode:1969Sci...165..698O. doi:10.1126/science.165.3894.698. PMID 5793973.
85. "Learning About Tay-Sachs Disease". U.S. National Human Genome Research Institute. Retrieved 1 March 2015.
86. Erlandsen H, Stevens RC (October 1999). "The structural basis of phenylketonuria". Molecular Genetics and Metabolism. 68 (2): 103–25. doi:10.1006/mgme.1999.2922. PMID 10527663.
87. Flatmark T, Stevens RC (August 1999). "Structural Insight into the Aromatic Amino Acid Hydroxylases and Their Disease-Related Mutant Forms". Chemical Reviews. 99 (8): 2137–2160. doi:10.1021/cr980450y. PMID 11849022.
88. «Phenylketonuria". Genes and Disease [Internet]. Bethesda (MD): National Center for Biotechnology Information (US). 1998–2015.
89. "Pseudocholinesterase deficiency". U.S. National Library of Medicine. Retrieved 5 September 2013.
90. Fieker A, Philpott J, Armand M (2011). "Enzyme replacement therapy for pancreatic insufficiency: present and future". Clinical and Experimental Gastroenterology. 4: 55–73. doi:10.2147/CEG.S17634. PMC 3132852. PMID 21753892.
91. Misselwitz B, Pohl D, Frühauf H, Fried M, Vavricka SR, Fox M (June 2013). "Lactose malabsorption and intolerance: pathogenesis, diagnosis and treatment". United European Gastroenterology Journal. 1 (3): 151–9. doi:10.1177/2050640613484463. PMC 4040760. PMID 24917953.
92. Cleaver JE (May 1968). "Defective repair replication of DNA in xeroderma pigmentosum". Nature. 218 (5142): 652–6. Bibcode:1968Natur.218..652C. doi:10.1038/218652a0. PMID 5655953.
93. James WD, Elston D, Berger TG (2011). Andrews' Diseases of the Skin: Clinical Dermatology (11th ed.). London: Saunders/ Elsevier. p. 567. ISBN 978-1437703146.
94. Renugopalakrishnan V, Garduño-Juárez R, Narasimhan G, Verma CS, Wei X, Li P (November 2005). "Rational design of thermally stable proteins: relevance to bionanotechnology". Journal of Nanoscience and Nanotechnology. 5 (11): 1759–1767. doi:10.1166/jnn.2005.441. PMID 16433409.
95. Hult K, Berglund P (August 2003). "Engineered enzymes for improved organic synthesis". Current Opinion in Biotechnology. 14 (4): 395–400. doi:10.1016/

S0958-1669(03)00095-8. PMID 12943848.

96. Jiang L, Althoff EA, Clemente FR, Doyle L, Röthlisberger D, Zanghellini A, Gallaher JL, Betker JL, Tanaka F, Barbas CF, Hilvert D, Houk KN, Stoddard BL, Baker D (March 2008). "De novo computational design of retro-aldol enzymes". Science. 319 (5868): 1387–91. Bibcode:2008Sci...319.1387J. doi:10.1126/science.1152692. PMC 3431203. PMID 18323453.

97. Sun Y, Cheng J (May 2002). "Hydrolysis of lignocellulosic materials for ethanol production: a review". Bioresource Technology. 83 (1): 1–11. doi:10.1016/S0960-8524(01)00212-7. PMID 12058826.

98. Kirk O, Borchert TV, Fuglsang CC (August 2002). "Industrial enzyme applications". Current Opinion in Biotechnology. 13 (4): 345–351. doi:10.1016/S0958-1669(02)00328-2. PMID 12323357.

99. Briggs DE (1998). Malts and Malting (1st ed.). London: Blackie Academic. ISBN 978-0412298004.

100. Dulieu C, Moll M, Boudrant J, Poncelet D (2000). "Improved performances and control of beer fermentation using encapsulated alpha-acetolactate decarboxylase and modeling". Biotechnology Progress. 16 (6): 958–65. doi:10.1021/bp000128k. PMID 11101321.

101. Tarté R (2008). Ingredients in Meat Products Properties, Functionality and Applications. New York: Springer. p. 177. ISBN 978-0-387-71327-4.

102. Chymosin – GMO Database". GMO Compass. European Union. 10 July 2010. Retrieved 1 March 2015.

103. Molimard P, Spinnler HE (February 1996). "Review: Compounds Involved in the Flavor of Surface Mold-Ripened Cheeses: Origins and Properties". Journal of Dairy Science. 79 (2): 169–184. doi:10.3168/jds.S0022-0302(96)76348-8.

104. Guzmán-Maldonado H, Paredes-López O (September 1995). "Amylolytic enzymes and products derived from starch: a review". Critical Reviews in Food Science and Nutrition. 35 (5): 373–403. doi:10.1080/10408399509527706. PMID 8573280.

105. "Protease – GMO Database". GMO Compass. European Union. 10 July 2010. Retrieved 28 February 2015.

106. Alkorta I, Garbisu C, Llama MJ, Serra JL (January 1998). "Industrial applications of pectic enzymes: a review". Process Biochemistry. 33 (1): 21–28. doi:10.1016/S0032-9592(97)00046-0.

107. Bajpai P (March 1999). "Application of enzymes in the pulp and paper industry". Biotechnology Progress. 15 (2): 147–157. doi:10.1021/bp990013k. PMID 10194388.

108. Begley CG, Paragina S, Sporn A (March 1990). "An analysis of contact lens enzyme cleaners". Journal of the American Optometric Association. 61 (3): 190–4. PMID 2186082.

109. Farris PL (2009). "Economic Growth and Organization of the U.S. Starch Industry". In BeMiller JN, Whistler RL. Starch Chemistry and Technology (3rd ed.). London: Academic. ISBN 9780080926551.

CHAPTER 2

Protein Structure and Functions

Chapter Contents

- Introduction
- Biochemistry
- Synthesis
- Structure
- Cellular Functions
- Methods of Study
- History and Etymology
- Protein Classification
- Secondary Structure
- Tertiary Structure
- Quaternary Structure
- Structural Alignment
- Macromolecular Docking
- Protein Structure

Proteins are large biomolecules, or macromolecules, consisting of one or more long chains of amino acid residues. Proteins perform a vast array of functions within organisms, including catalysing metabolic reactions, DNA replication, responding to stimuli, and transporting molecules from one location to another. Proteins differ from one another primarily in their sequence of amino acids, which is dictated by the nucleotide sequence of their genes, and which usually results in protein folding into a specific three-dimensional structure that determines its activity.

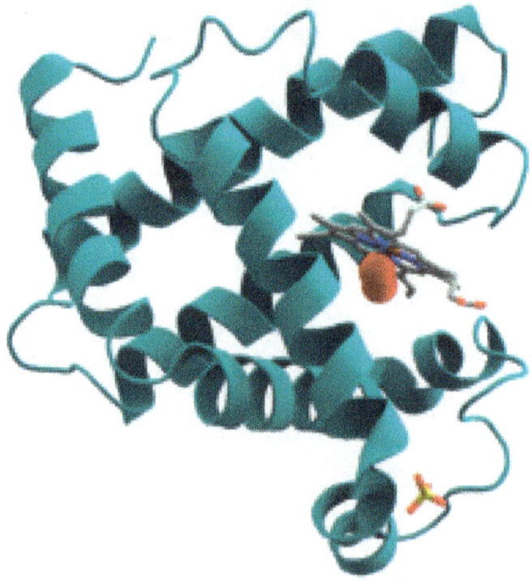

A representation of the 3D structure of the protein myoglobin showing turquoise α-helices. This protein was the first to have its structure solved by X-ray crystallography. Towards the right-center among the coils, a prosthetic group called a heme group (shown in gray) with a bound oxygen molecule (red).

INTRODUCTION

A linear chain of amino acid residues is called a polypeptide. A protein contains at least one long polypeptide. Short polypeptides, containing less than 20–30 residues, are rarely considered to be proteins and are commonly called peptides, or sometimes oligopeptides. The individual amino acid residues are bonded together by peptide bonds and adjacent amino acid residues. The sequence of amino acid residues in a protein is defined by the sequence of a gene, which is encoded in the genetic code. In general, the genetic code specifies 20 standard amino acids; however, in certain organisms the genetic code can include selenocysteine and—in certain archaea—pyrrolysine. Shortly after or even during synthesis, the residues in a protein are often chemically modified by post-translational modification, which alters the physical and chemical properties, folding, stability, activity, and ultimately, the function of the proteins. Sometimes proteins have non-peptide groups attached, which can be called prosthetic groups or cofactors. Proteins can also work together to achieve a particular function, and they often associate to form stable protein complexes.

Once formed, proteins only exist for a certain period of time and are then degraded and recycled by the cell's machinery through the process of protein turnover. A protein's lifespan is measured in terms of its half-life and covers a wide range. They can exist for minutes or years with an average lifespan of 1–2 days in mammalian cells. Abnormal and or misfolded proteins are degraded more rapidly either due to being targeted for destruction or due to being unstable.

Like other biological macromolecules such as polysaccharides and nucleic acids, proteins are essential parts of organisms and participate in virtually every process within cells. Many proteins are enzymes that catalyse biochemical reactions and are vital to metabolism. Proteins also have structural or mechanical functions, such as actin and myosin in muscle and the proteins in the cytoskeleton, which form a system of scaffolding that maintains cell shape. Other proteins are important in cell signaling, immune responses, cell adhesion, and the cell cycle. In animals, proteins are needed in the diet to provide the essential amino acids that cannot be synthesized. Digestion breaks the proteins down for use in the metabolism.

Proteins may be purified from other cellular components using a variety of techniques such as ultracentrifugation, precipitation, electrophoresis, and chromatography; the advent of genetic engineering has made possible a number of methods to facilitate purification. Methods commonly used to study protein structure and function include immunohistochemistry, site-directed mutagenesis, X-ray crystallography, nuclear magnetic resonance and mass spectrometry.

BIOCHEMISTRY

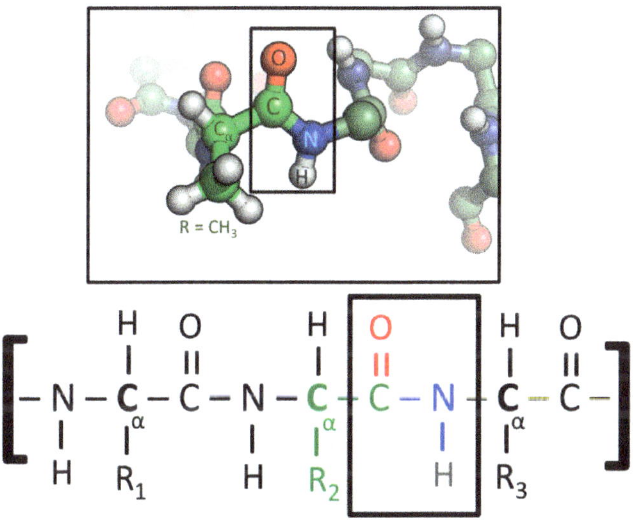

Chemical structure of the peptide bond (bottom) and the three-dimensional structure of a peptide bond between an alanine and an adjacent amino acid (top/inset).

Resonance structures of the peptide bond that links individual amino acids to form a protein polymer.

Most proteins consist of linear polymers built from series of up to 20 different L-α-amino acids. All proteinogenic amino acids possess common structural features, including an α-carbon to which an amino group, a carboxyl group, and a variable side chain are bonded. Only proline differs from this basic structure as it contains an unusual ring to the N-end amine group, which forces the CO–NH amide moiety into a fixed conformation. The side chains of the standard amino acids, detailed in the list of standard amino acids, have a great variety of chemical structures and properties; it is the combined effect of all of the amino acid side chains in a protein that ultimately determines its three-dimensional structure and its chemical reactivity. The amino acids in a polypeptide chain are linked by peptide bonds. Once linked in the protein chain, an individual amino acid is called a residue, and the linked series of carbon, nitrogen, and oxygen atoms are known as the main chain or protein backbone.

The peptide bond has two resonance forms that contribute some double-bond character and inhibit rotation around its axis, so that the alpha carbons are roughly coplanar. The other two dihedral angles in the peptide bond determine the local shape assumed by the protein backbone. The end with a free amino group is known as the N-terminus or amino terminus, whereas the end of the protein with a free carboxyl group is known as the C-terminus or carboxy terminus (the sequence of the protein is written from N-terminus to C-terminus, from left to right).

The words protein, polypeptide, and peptide are a little ambiguous and can overlap in meaning. Protein is generally used to refer to the complete biological molecule in a stable conformation, whereas peptide is generally reserved for a short amino acid oligomers often lacking a stable three-dimensional structure. However, the boundary between the two is not well defined and usually lies near 20–30 residues. Polypeptide can refer to any single linear chain of amino acids, usually regardless of length, but often implies an absence of a defined conformation.

Abundance in Cells

It has been estimated that average-sized bacteria contain about 2 million proteins per cell (e.g. E. coli and Staphylococcus aureus). Smaller bacteria, such as Mycoplasma or spirochetes contain fewer molecules, namely on the order of 50,000 to 1 million. By contrast, eukaryotic cells are larger and thus contain much more protein. For instance, yeast cells were estimated to contain about 50 million proteins and human cells on the order of 1 to 3 billion. The concentration of individual protein copies ranges from a few molecules per cell up to 20 million.Not all genes coding proteins are expressed in most cells and their number depends on for example cell type and external stimuli. For instance, of the 20,000 or so proteins encoded by the human genome, only 6,000 are detected in lymphoblastoid cells. Moreover, the number of proteins the genome encodes correlates well with the organism complexity. Eukaryotes, bacteria, Archaea and viruses have on average 15145, 3200, 2358 and 42 proteins respectively coded in their genomes.

SYNTHESIS

Biosynthesis

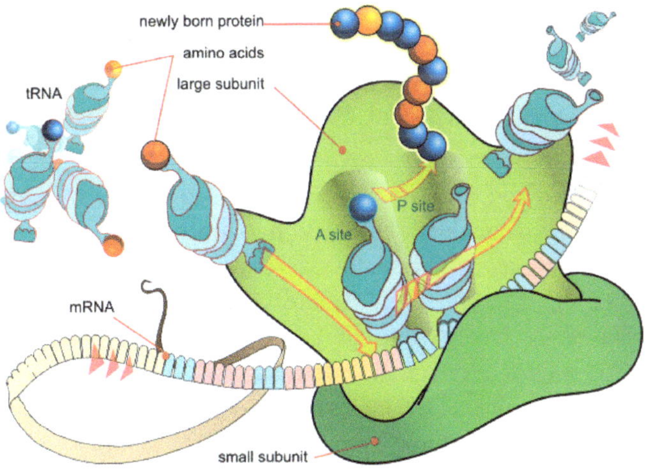

A ribosome produces a protein using mRNA as template.

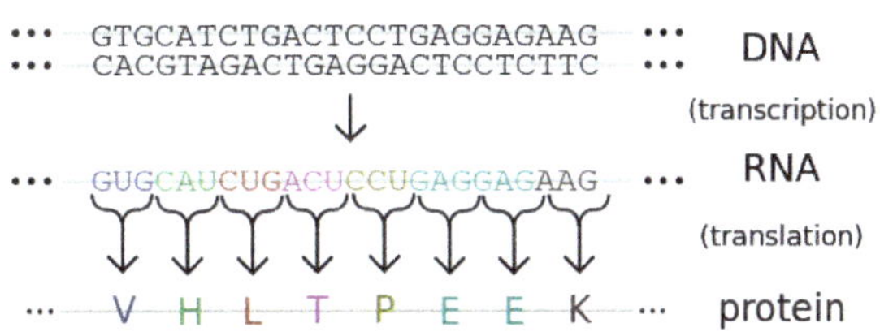

The DNA sequence of a gene encodes the amino acid sequence of a protein.

Proteins are assembled from amino acids using information encoded in genes. Each protein has its own unique amino acid sequence that is specified by the nucleotide sequence of the gene encoding this protein. The genetic code is a set of three-nucleotide sets called codons and each three-nucleotide combination designates an amino acid, for example AUG (adenine-uracil-guanine) is the code for methionine. Because DNA contains four nucleotides, the total number of possible codons is 64; hence, there is some redundancy in the genetic code, with some amino acids specified by more than one codon. Genes encoded in DNA are first transcribed into pre-messenger RNA (mRNA) by proteins such as RNA polymerase. Most organisms then process the pre-mRNA (also known as a primary transcript) using various forms of Post-transcriptional modification to form the mature mRNA, which is then used as a template for protein synthesis by the ribosome. In prokaryotes the mRNA may either be used as soon as it is produced, or be bound by a ribosome after having moved away from the nucleoid. In contrast, eukaryotes make mRNA in the cell nucleus and then translocate it across the nuclear membrane into the cytoplasm, where protein synthesis then takes place. The rate of protein synthesis is higher in prokaryotes than eukaryotes and can reach up to 20 amino acids per second.

The process of synthesizing a protein from an mRNA template is known as translation. The mRNA is loaded onto the ribosome and is read three nucleotides at a time by matching each codon to its base pairing anticodon located on a transfer RNA molecule, which carries the amino acid corresponding to the codon it recognizes. The enzyme aminoacyl tRNA synthetase "charges" the tRNA molecules with the correct amino acids. The growing polypeptide is often termed the nascent chain. Proteins are always biosynthesized from N-terminus to C-terminus.

The size of a synthesized protein can be measured by the number of amino acids it contains and by its total molecular mass, which is normally reported in units of daltons (synonymous with atomic mass units), or the derivative unit kilodalton (kDa). The average size of protein increases from Archaea, Bacteria to Eukaryote (283, 311, 438 residues and 31, 34, 49 kDa respecitvely) due bigger number of protein domains constituting proteins in higher organisms. For instance, yeast proteins are on average 466 amino acids long and 53 kDa in mass. The largest known proteins are the titins, a component of the muscle sarcomere, with a molecular mass of almost 3,000 kDa and a total length of almost 27,000 amino acids.

Chemical Synthesis

Short proteins can also be synthesized chemically by a family of methods known as peptide synthesis, which rely on organic synthesis techniques such as chemical ligation to produce peptides in high yield. Chemical synthesis allows for the introduction of non-natural amino acids into polypeptide chains, such as attachment of fluorescent probes to amino acid side chains. These methods are useful in laboratory biochemistry and cell biology, though generally not for commercial applications. Chemical synthesis is inefficient for polypeptides longer than about 300 amino acids, and the synthesized proteins may not readily assume their native tertiary structure. Most chemical synthesis methods proceed from C-terminus to N-terminus, opposite the biological reaction.

STRUCTURE

The crystal structure of the chaperonin, a huge protein complex. A single protein subunit is highlighted. Chaperonins assist protein folding.

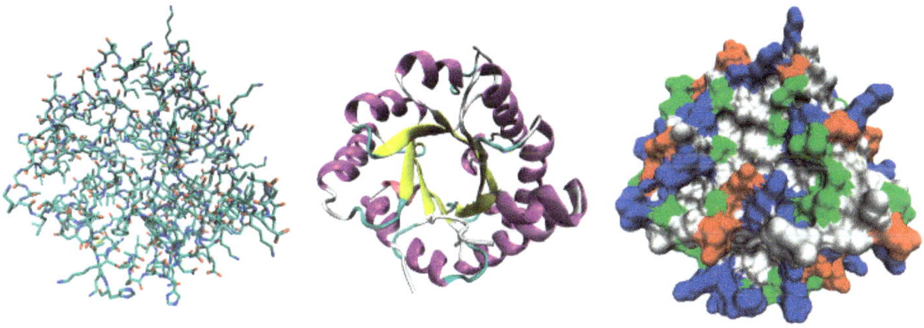

Three possible representations of the three-dimensional structure of the protein triose phosphate isomerase. **Left**: All-atom representation colored by atom type. **Middle:** Simplified representation illustrating the backbone conformation, colored by secondary structure. **Right**: Solvent-accessible surface representation colored by residue type (acidic residues red, basic residues blue, polar residues green, nonpolar residues white).

Most proteins fold into unique 3-dimensional structures. The shape into which a protein naturally folds is known as its native conformation. Although many proteins can fold unassisted, simply through the chemical properties of their amino acids, others require the aid of molecular chaperones to fold into their native states. Biochemists often refer to four distinct aspects of a protein's structure:

- *Primary structure*: the amino acid sequence. A protein is a polyamide.
- *Secondary structure*: regularly repeating local structures stabilized by hydrogen bonds. The most common examples are the α-helix, β-sheet and turns. Because secondary structures are local, many regions of different secondary structure can be present in the same protein molecule.
- *Tertiary structure*: the overall shape of a single protein molecule; the spatial relationship of the secondary structures to one another. Tertiary structure is generally stabilized by nonlocal interactions, most commonly the formation of a hydrophobic core, but also through salt bridges, hydrogen bonds, disulfide bonds, and even posttranslational modifications. The term "tertiary structure" is often used as synonymous with the term *fold*. The tertiary structure is what controls the basic function of the protein.
- *Quaternary structure*: the structure formed by several protein molecules (polypeptide chains), usually called *protein subunits* in this context, which function as a single protein complex.

Proteins are not entirely rigid molecules. In addition to these levels of structure, proteins may shift between several related structures while they perform their functions. In the context of these functional rearrangements, these tertiary or quaternary structures are usually referred to as "conformations", and transitions between them are called conformational changes. Such changes are often induced by the binding of a substrate molecule to an enzyme's active site, or the physical region of the protein that participates in chemical catalysis. In solution proteins also undergo variation in structure through thermal vibration and the collision with other molecules.

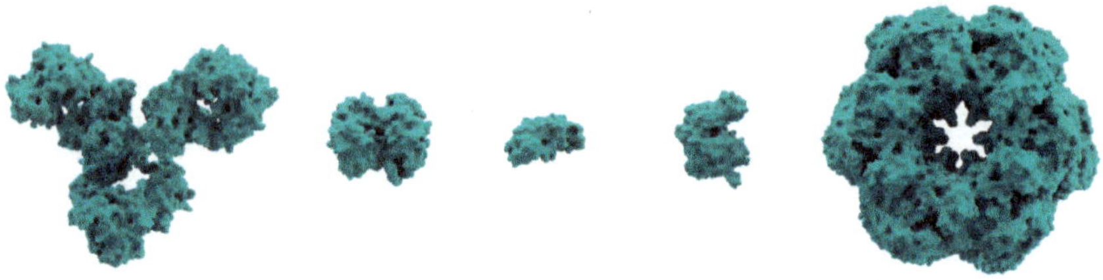

Molecular surface of several proteins showing their comparative sizes. From left to right are: immunoglobulin G (IgG, an antibody), hemoglobin, insulin (a hormone), adenylate kinase (an enzyme), and glutamine synthetase (an enzyme).

Proteins can be informally divided into three main classes, which correlate with typical tertiary structures: globular proteins, fibrous proteins, and membrane proteins. Almost all globular proteins are soluble and many are enzymes. Fibrous proteins are often structural, such as collagen, the major component of connective tissue, or keratin, the protein component of hair and nails. Membrane proteins often serve as receptors or provide channels for polar or charged molecules to pass through the cell membrane.

A special case of intramolecular hydrogen bonds within proteins, poorly shielded from water attack and hence promoting their own dehydration, are called dehydrons.

Structure Determination

Discovering the tertiary structure of a protein, or the quaternary structure of its complexes, can provide important clues about how the protein performs its function. Common experimental methods of structure determination include X-ray crystallography and NMR spectroscopy, both of which can produce information at atomic resolution. However, NMR experiments are able to provide information from which a subset of distances between pairs of atoms can be estimated, and the final possible conformations for a protein are determined by solving a distance geometry problem. Dual polarisation interferometry is a quantitative analytical method for measuring the overall protein conformation and conformational changes due to interactions or other stimulus. Circular dichroism is another laboratory technique for determining internal β-sheet / α-helical composition of proteins. Cryoelectron microscopy is used to produce lower-resolution structural information about very large protein complexes, including assembled viruses; a variant known as electron crystallography can also produce high-resolution information in some cases, especially for two-dimensional crystals of membrane proteins. Solved structures are usually deposited in the Protein Data Bank (PDB), a freely available resource from which structural data about thousands of proteins can be obtained in the form of Cartesian coordinates for each atom in the protein.

Many more gene sequences are known than protein structures. Further, the set of solved structures is biased toward proteins that can be easily subjected to the conditions required in X-ray crystallography, one of the major structure determination methods. In particular, globular proteins are comparatively easy to crystallize in preparation for X-ray crystallography. Membrane proteins, by contrast, are difficult to crystallize and are underrepresented in the PDB. Structural genomics initiatives have attempted to remedy these deficiencies by systematically solving representative structures of major fold classes. Protein structure prediction methods attempt to provide a means of generating a plausible structure for proteins whose structures have not been experimentally determined.

CELLULAR FUNCTIONS

Proteins are the chief actors within the cell, said to be carrying out the duties specified by the information encoded in genes. With the exception of certain types of RNA, most other biological molecules are relatively inert elements upon which proteins act. Proteins make up half the dry weight of an Escherichia coli cell, whereas other macromolecules such as DNA and RNA make up only 3% and 20%, respectively. The set of proteins expressed in a particular cell or cell type is known as its proteome.

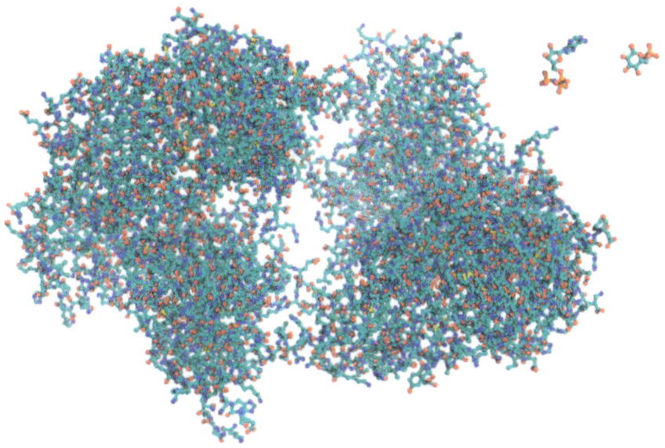

The enzyme hexokinase is shown as a conventional ball-and-stick molecular model. To scale in the top right-hand corner are two of its substrates, ATP and glucose.

The chief characteristic of proteins that also allows their diverse set of functions is their ability to bind other molecules specifically and tightly. The region of the protein responsible for binding another molecule is known as the binding site and is often a depression or "pocket" on the molecular surface. This binding ability is mediated by the tertiary structure of the protein, which defines the binding site pocket, and by the chemical properties of the surrounding amino acids' side chains. Protein binding can be extraordinarily tight and specific; for example, the ribonuclease inhibitor protein binds to human angiogenin with a sub-femtomolar dissociation constant (<10−15 M) but does not bind at all to its amphibian homolog onconase (>1 M). Extremely minor chemical changes such as the addition of a single methyl group to a binding partner can sometimes suffice to nearly eliminate binding; for example, the aminoacyl tRNA synthetase specific to the amino acid valine discriminates against the very similar side chain of the amino acid isoleucine.

Proteins can bind to other proteins as well as to small-molecule substrates. When proteins bind specifically to other copies of the same molecule, they can oligomerize to form fibrils; this process occurs often in structural proteins that consist of globular monomers that self-associate to form rigid fibers. Protein–protein interactions also regulate enzymatic activity, control progression through the cell cycle, and allow the assembly of large protein complexes that carry out many closely related reactions with a common biological function. Proteins can also bind to, or even be integrated into, cell membranes. The ability of binding partners to induce conformational changes in proteins allows the construction of enormously complex signaling networks. As interactions between proteins are reversible, and depend heavily on the availability of different groups of partner proteins to form aggregates that are capable to carry out discrete sets of function, study of the interactions between specific proteins is a key to understand important aspects of cellular function, and ultimately the properties that distinguish particular cell types.

Enzymes

The best-known role of proteins in the cell is as enzymes, which catalyse chemical reactions. Enzymes are usually highly specific and accelerate only one or a few chemical reactions. Enzymes carry out most of the reactions involved in metabolism, as well as manipulating DNA in processes such as DNA replication, DNA repair, and transcription. Some enzymes act on other proteins to add or remove chemical groups in a process known as posttranslational modification. About 4,000 reactions are known to be catalysed by enzymes. The rate acceleration conferred by enzymatic catalysis is often enormous—as much as 1017-fold increase in rate over the uncatalysed reaction in the case of orotate decarboxylase (78 million years without the enzyme, 18 milliseconds with the enzyme).

The molecules bound and acted upon by enzymes are called substrates. Although enzymes can consist of hundreds of amino acids, it is usually only a small fraction of the residues that come in contact with the substrate, and an even smaller fraction—three to four residues on average—that are directly involved in catalysis. The region of the enzyme that binds the substrate and contains the catalytic residues is known as the active site.

Dirigent proteins are members of a class of proteins that dictate the stereochemistry of a compound synthesized by other enzymes.

Cell Signaling and Ligand Binding

Many proteins are involved in the process of cell signaling and signal transduction. Some proteins, such as insulin, are extracellular proteins that transmit a signal from the cell in which they were synthesized to other cells in distant tissues. Others are membrane proteins that act as receptors whose main function is to bind a signaling molecule and induce a biochemical response in the cell. Many receptors have a binding site exposed on the cell surface and an effector domain within the cell, which may have enzymatic activity or may undergo a conformational change detected by other proteins within the cell.

Ribbon diagram of a mouse antibody against cholera that binds a carbohydrate antigen.

Antibodies are protein components of an adaptive immune system whose main function is to bind antigens, or foreign substances in the body, and target them for destruction. Antibodies can be secreted into the extracellular environment or anchored in the membranes of specialized B cells known as plasma cells. Whereas enzymes are limited in their binding affinity for their substrates by the necessity of conducting their reaction, antibodies have no such constraints. An antibody's binding affinity to its target is extraordinarily high.

Many ligand transport proteins bind particular small biomolecules and transport them to other locations in the body of a multicellular organism. These proteins must have a high binding affinity when their ligand is present in high concentrations, but must also release the ligand when it is present at low concentrations in the target tissues. The canonical example of a ligand-binding protein is haemoglobin, which transports oxygen from the lungs to other organs and tissues in all vertebrates and has close homologs in every biological kingdom. Lectins are sugar-binding proteins which are highly specific for their sugar moieties. Lectins typically play a role in biological recognition phenomena involving cells and proteins. Receptors and hormones are highly specific binding proteins.

Transmembrane proteins can also serve as ligand transport proteins that alter the permeability of the cell membrane to small molecules and ions. The membrane alone has a hydrophobic core through which polar or charged molecules cannot diffuse. Membrane proteins contain internal channels that allow such molecules to enter and exit the cell. Many ion channel proteins are specialized to select for only a particular ion; for example, potassium and sodium channels often discriminate for only one of the two ions.

Structural Proteins

Structural proteins confer stiffness and rigidity to otherwise-fluid biological components. Most structural proteins are fibrous proteins; for example, collagen and elastin are critical components of connective tissue such as cartilage, and keratin is found in hard or filamentous structures such as hair, nails, feathers, hooves, and some animal shells. Some globular proteins can also play structural functions, for example, actin and tubulin are globular and soluble as monomers, but polymerize to form long, stiff fibers that make up the cytoskeleton, which allows the cell to maintain its shape and size.

Other proteins that serve structural functions are motor proteins such as myosin, kinesin, and dynein, which are capable of generating mechanical forces. These proteins are crucial for cellular motility of single celled organisms and the sperm of many multicellular organisms which reproduce sexually. They also generate the forces exerted by contracting muscles and play essential roles in intracellular transport.

METHODS OF STUDY

The activities and structures of proteins may be examined in vitro, in vivo, and in silico. In vitro studies of purified proteins in controlled environments are useful for learning how a protein carries out its function: for example, enzyme kinetics studies explore the chemical mechanism of an enzyme's catalytic activity and its relative affinity for various possible substrate molecules. By contrast, in vivo experiments can provide information about the physiological role of a protein in the context of a cell or even a whole organism. In silico studies use computational methods to study proteins.

Protein Purification

To perform in vitro analysis, a protein must be purified away from other cellular components. This process usually begins with cell lysis, in which a cell's membrane is disrupted and its internal contents released into a solution known as a crude lysate. The resulting mixture can be purified using ultracentrifugation, which fractionates the various cellular components into fractions containing soluble proteins; membrane lipids and proteins; cellular organelles, and nucleic acids. Precipitation by a method known as salting out can concentrate the proteins from this lysate. Various types of chromatography are then used to isolate the protein or proteins of interest based on properties such as molecular weight, net charge and binding affinity.The level of purification can be monitored using various types of gel electrophoresis if the desired protein's molecular weight and isoelectric point are known, by spectroscopy if the protein has distinguishable spectroscopic features, or by enzyme assays if the protein has enzymatic activity. Additionally, proteins can be isolated according their charge using electrofocusing.

For natural proteins, a series of purification steps may be necessary to obtain protein sufficiently pure for laboratory applications. To simplify this process, genetic engineering is often used to add chemical features to proteins that make them easier to purify without affecting their structure or activity. Here, a "tag" consisting of a specific amino acid sequence, often a series of histidine residues (a "His-tag"), is attached to one terminus of the protein. As a result, when the lysate is passed over a chromatography column containing nickel, the histidine residues ligate the nickel and attach to the column while the untagged components of the lysate pass unimpeded. A number of different tags have been developed to help researchers purify specific proteins from complex mixtures.

Cellular Localization

The study of proteins in vivo is often concerned with the synthesis and localization of the protein within the cell. Although many intracellular proteins are synthesized in the cytoplasm and membrane-bound or secreted proteins in the endoplasmic reticulum, the specifics of how proteins are targeted to specific organelles or cellular structures is often unclear. A useful technique for assessing cellular localization uses genetic engineering to express in a cell a fusion protein or chimera consisting of the natural protein of interest linked to a "reporter" such as green fluorescent protein (GFP). The fused protein's position within the cell can be cleanly and efficiently visualized using microscopy, as shown in the figure opposite.

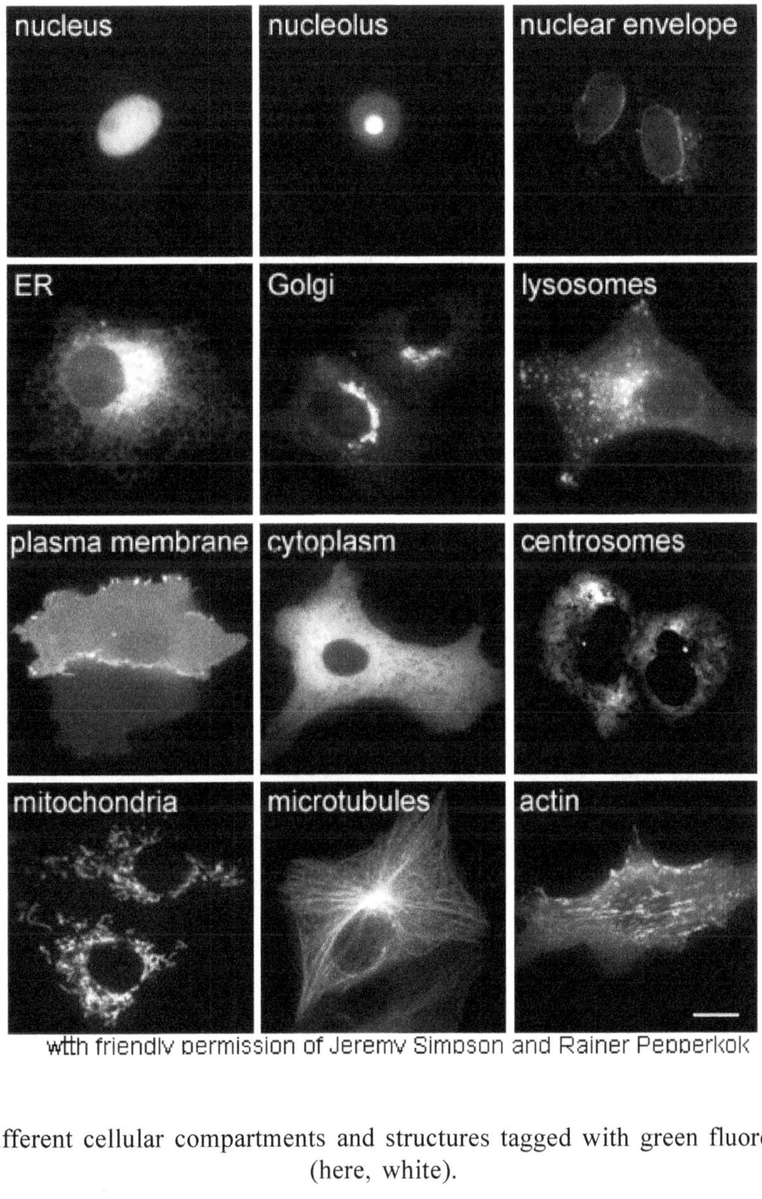

wtth friendlv permission of Jeremv Simpson and Rainer Pepperkok

Proteins in different cellular compartments and structures tagged with green fluorescent protein (here, white).

Other methods for elucidating the cellular location of proteins requires the use of known compartmental markers for regions such as the ER, the Golgi, lysosomes or vacuoles, mitochondria, chloroplasts, plasma membrane, etc. With the use of fluorescently tagged versions of these markers or of antibodies to known markers, it becomes much simpler to identify the localization of a protein of interest. For example, indirect immunofluorescence will allow for fluorescence colocalization and demonstration of location. Fluorescent dyes are used to label cellular compartments for a similar purpose.

Other possibilities exist, as well. For example, immunohistochemistry usually utilizes an antibody to one or more proteins of interest that are conjugated to enzymes yielding either luminescent or chromogenic signals that can be compared between samples, allowing for localization information. Another applicable technique is cofractionation in sucrose (or other material) gradients using isopycnic centrifugation. While this technique does not prove colocalization of a compartment of known density and the protein of interest, it does increase the likelihood, and is more amenable to large-scale studies.

Finally, the gold-standard method of cellular localization is immunoelectron microscopy. This technique also uses an antibody to the protein of interest, along with classical electron microscopy techniques. The sample is prepared for normal electron microscopic examination, and then treated with an antibody to the protein of interest that is conjugated to an extremely electro-dense material, usually gold. This allows for the localization of both ultrastructural details as well as the protein of interest.

Through another genetic engineering application known as site-directed mutagenesis, researchers can alter the protein sequence and hence its structure, cellular localization, and susceptibility to regulation. This technique even allows the incorporation of unnatural amino acids into proteins, using modified tRNAs, and may allow the rational design of new proteins with novel properties.

Proteomics

The total complement of proteins present at a time in a cell or cell type is known as its proteome, and the study of such large-scale data sets defines the field of proteomics, named by analogy to the related field of genomics. Key experimental techniques in proteomics include 2D electrophoresis, which allows the separation of a large number of proteins, mass spectrometry, which allows rapid high-throughput identification of proteins and sequencing of peptides (most often after in-gel digestion), protein microarrays, which allow the detection of the relative levels of a large number of proteins present in a cell, and two-hybrid screening, which allows the systematic exploration of protein–protein interactions. The total complement of biologically possible such interactions is known as the interactome. A systematic attempt to determine the structures of proteins representing every possible fold is known as structural genomics.

Bioinformatics

A vast array of computational methods have been developed to analyze the structure, function, and evolution of proteins.

The development of such tools has been driven by the large amount of genomic and proteomic data available for a variety of organisms, including the human genome. It is simply impossible to study all proteins experimentally, hence only a few are subjected to laboratory experiments while computational tools are used to extrapolate to similar proteins. Such homologous proteins can be efficiently identified in distantly related organisms by sequence alignment. Genome and gene sequences can be searched by a variety of tools for certain properties. Sequence profiling tools can find restriction enzyme sites, open reading frames in nucleotide sequences, and predict secondary structures. Phylogenetic trees can be constructed and evolutionary hypotheses developed using special software like ClustalW regarding the ancestry of modern organisms and the genes they express. The field of bioinformatics is now indispensable for the analysis of genes and proteins.

Structure Prediction and Simulation

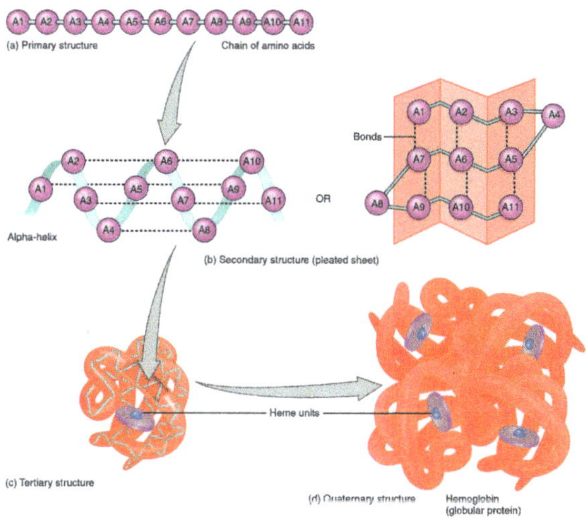

Constituent amino-acids can be analyzed to predict secondary, tertiary and quaternary protein structure, in this case hemoglobin containing heme units.

Complementary to the field of structural genomics, protein structure prediction seeks to develop efficient ways to provide plausible models for proteins whose structures have not yet been determined experimentally. The most successful type of structure prediction, known as homology modeling, relies on the existence of a "template" structure with sequence similarity to the protein being modeled; structural genomics' goal is to provide sufficient representation in solved structures to model most of those that remain. Although producing accurate models remains a challenge when only distantly related template structures are available, it has been suggested that sequence alignment is the bottleneck in this process, as quite accurate models can be produced if a "perfect" sequence alignment is known. Many structure prediction methods have served to inform the emerging field of protein engineering, in which novel protein folds have already been designed. A more complex computational problem is the prediction of intermolecular interactions, such as in molecular docking and protein–protein interaction prediction.

The processes of protein folding and binding can be simulated using such technique as molecular mechanics, in particular, molecular dynamics and Monte Carlo, which increasingly take advantage of parallel and distributed computing (Folding@home project; molecular modeling on GPU). The folding of small α-helical protein domains such as the villin headpiece and the HIV accessory protein have been successfully simulated in silico, and hybrid methods that combine standard molecular dynamics with quantum mechanics calculations have allowed exploration of the electronic states of rhodopsins.

Protein Disorder and Unstructure Prediction

Many proteins (in Eucaryota ~33%) contain large unstructured but biologically functional segments and can be classified as intrinsically disordered proteins. Predicting and analysing protein disorder is, therefore, an important part of protein structure characterisation.

Nutrition

Most microorganisms and plants can biosynthesize all 20 standard amino acids, while animals (including humans) must obtain some of the amino acids from the diet. The amino acids that an organism cannot synthesize on its own are referred to as essential amino acids. Key enzymes that synthesize certain amino acids are not present in animals — such as aspartokinase, which catalyses the first step in the synthesis of lysine, methionine, and threonine from aspartate. If amino acids are present in the environment, microorganisms can conserve energy by taking up the amino acids from their surroundings and downregulating their biosynthetic pathways.

In animals, amino acids are obtained through the consumption of foods containing protein. Ingested proteins are then broken down into amino acids through digestion, which typically involves denaturation of the protein through exposure to acid and hydrolysis by enzymes called proteases. Some ingested amino acids are used for protein biosynthesis, while others are converted to glucose through gluconeogenesis, or fed into the citric acid cycle. This use of protein as a fuel is particularly important under starvation conditions as it allows the body's own proteins to be used to support life, particularly those found in muscle. Amino acids are also an important dietary source of nitrogen.

HISTORY AND ETYMOLOGY

Proteins were recognized as a distinct class of biological molecules in the eighteenth century by Antoine Fourcroy and others, distinguished by the molecules' ability to coagulate or flocculate under treatments with heat or acid. Noted examples at the time included albumin from egg whites, blood serum albumin, fibrin, and wheat gluten.

Proteins were first described by the Dutch chemist Gerardus Johannes Mulder and named by the Swedish chemist Jöns Jacob Berzelius in 1838. Mulder carried out elemental analysis of common proteins and found that nearly all proteins had the same empirical formula, C400H620N100O120P1S1. He came to the erroneous conclusion that they might be composed of a single type of (very large) molecule. The term "protein" to describe these molecules was proposed by Mulder's associate Berzelius; protein is derived from the Greek word πρώτειος (proteios), meaning "primary", "in the lead", or "standing in front", + -in. Mulder went on to identify the products of protein degradation such as the amino acid leucine for which he found a (nearly correct) molecular weight of 131 Da.

Early nutritional scientists such as the German Carl von Voit believed that protein was the most important nutrient for maintaining the structure of the body, because it was generally believed that "flesh makes flesh." Karl Heinrich Ritthausen extended known protein forms with the identification of glutamic acid. At the Connecticut Agricultural Experiment Station a detailed review of the vegetable proteins was compiled by Thomas Burr Osborne. Working with Lafayette Mendel and applying Liebig's law of the minimum in feeding laboratory rats, the nutritionally essential amino acids were established. The work was continued and communicated by William Cumming Rose. The understanding of proteins as polypeptides came through the work of Franz Hofmeister and Hermann Emil Fischer. The central role of proteins as enzymes in living organisms was not fully appreciated until 1926, when James B. Sumner showed that the enzyme urease was in fact a protein.

The difficulty in purifying proteins in large quantities made them very difficult for early protein biochemists to study. Hence, early studies focused on proteins that could be purified in large quantities, e.g., those of blood, egg white, various toxins, and digestive/metabolic enzymes obtained from slaughterhouses. In the 1950s, the Armour Hot Dog Co. purified 1 kg of pure bovine pancreatic ribonuclease A and made it freely available to scientists; this gesture helped ribonuclease A become a major target for biochemical study for the following decades.

John Kendrew with model of myoglobin in progress.

Linus Pauling is credited with the successful prediction of regular protein secondary structures based on hydrogen bonding, an idea first put forth by William Astbury in 1933. Later work by Walter Kauzmann on denaturation, based partly on previous studies by Kaj Linderstrøm-Lang, contributed an understanding of protein folding and structure mediated by hydrophobic interactions.

The first protein to be sequenced was insulin, by Frederick Sanger, in 1949. Sanger correctly determined the amino acid sequence of insulin, thus conclusively demonstrating that proteins consisted of linear polymers of amino acids rather than branched chains, colloids, or cyclols.He won the Nobel Prize for this achievement in 1958.

The first protein structures to be solved were hemoglobin and myoglobin, by Max Perutz and Sir John Cowdery Kendrew, respectively, in 1958. As of 2017, the Protein Data Bank has over 126,060 atomic-resolution structures of proteins. In more recent times, cryo-electron microscopy of large macromolecular assemblies and computational protein structure prediction of small protein domains are two methods approaching atomic resolution.

PROTEIN CLASSIFICATION

Proteins may be classified according to both structural and sequence similarity. For structural classification, the sizes and spatial arrangements of secondary structures described in the above paragraph are compared in known three-dimensional structures.Classification based on sequence similarity was historically the first to be used. Initially, similarity based on alignments of whole sequences was performed. Later, proteins were classified on the basis of the occurrence of conserved amino acid patterns. Databases that classify proteins by one or more of these schemes are available. In considering protein classification schemes, it is important to keep several observations in mind. First, two entirely different protein sequences from different evolutionary origins may fold into a similar structure. Conversely, the sequence of an ancient gene for a given structure may have diverged considerably in different species while at the same time maintaining the same basic structural features. Recognizing any remaining sequence similarity in such cases may be a very difficult task. Second, two proteins that share a significant degree of sequence similarity either with each other or with a third sequence also share an evolutionary origin and should share some structural features also. However, gene duplication and genetic rearrangements during evolution may give rise to new gene copies, which can then evolve into proteins with new function and structure.

Terms Used for Classifying Protein Structures and Sequences

The more commonly used terms for evolutionary and structural relationships among proteins are listed below. Many additional terms are used for various kinds of structural features found in proteins. Descriptions of such terms may be found at the CATH Web site the Structural Classification of Proteins (SCOP) Web site and a Glaxo-Wellcome tutorial on the Swiss bioinformatics Expasy Web site.

Active Site

A localized combination of amino acid side groups within the tertiary (three-dimensional) or quaternary (protein subunit) structure that can interact with a chemically specific substrate and that provides the protein with biological activity. Proteins of very different amino acid sequences may fold into a structure that produces the same active site.

Architecture

The relative orientations of secondary structures in a three-dimensional structure without regard to whether or not they share a similar loop structure.

Fold

A type of architecture that also has a conserved loop structure.

Blocks

A conserved amino acid sequence pattern in a family of proteins. The pattern includes a series of possible matches at each position in the represented sequences, but there are not any inserted or deleted positions in the pattern or in the sequences. By way of contrast, sequence profiles are a type of scoring matrix that represents a similar set of patterns that includes insertions and deletions.

Class

A term used to classify protein domains according to their secondary structural content and organization. Four classes were originally recognized by Levitt and Chothia (1976), and several others have been added in the SCOP database. Three classes are given in the CATH database: mainly-α, mainly-β, and α–β, with the α–β class including both alternating α/β and α+β structures.

Core

The portion of a folded protein molecule that comprises the hydrophobic interior of α-helices and β-sheets. The compact structure brings together side groups of amino acids into close enough proximity so that they can interact. When comparing protein structures, as in the SCOP database, core is the region common to most of the structures that share a common fold or that are in the same superfamily. In structure prediction, core is sometimes defined as the arrangement of secondary structures that is likely to be conserved during evolutionary change.

Domain (Sequence Context)

A segment of a polypeptide chain that can fold into a three-dimensional structure irrespective of the presence of other segments of the chain. The separate domains of a given protein may interact extensively or may be joined only by a length of polypeptide chain. A protein with several domains may use these domains for functional interactions with different molecules.

Family (Sequence Context)

A group of proteins of similar biochemical function that are more than 50% identical when aligned. This same cutoff is still used by the Protein Information Resource (PIR).

A protein family comprises proteins with the same function in different organisms (orthologous sequences) but may also include proteins in the same organism (paralogous sequences) derived from gene duplication and rearrangements. If a multiple sequence alignment of a protein family reveals a common level of similarity throughout the lengths of the proteins, PIR refers to the family as a homeomorphic family. The aligned region is referred to as a homeomorphic domain, and this region may comprise several smaller homology domains that are shared with other families. Families may be further subdivided into subfamilies or grouped into superfamilies based on respective higher or lower levels of sequence similarity. The SCOP database reports 1296 families and the CATH database (version 1.7 beta), reports 1846 families.

When the sequences of proteins with the same function are examined in greater detail, some are found to share high sequence similarity. They are obviously members of the same family by the above criteria. However, others are found that have very little, or even insignificant, sequence similarity with other family members. In such cases, the family relationship between two distant family members A and C can often be demonstrated by finding an additional family member B that shares significant similarity with both A and C. Thus, B provides a connecting link between A and C. Another approach is to examine distant alignments for highly conserved matches.

At a level of identity of 50%, proteins are likely to have the same three-dimensional structure, and the identical atoms in the sequence alignment will also superimpose within approximately 1 Å in the structural model. Thus, if the structure of one member of a family is known, a reliable prediction may be made for a second member of the family, and the higher the identity level, the more reliable the prediction. Protein structural modeling can be performed by examining how well the amino acid substitutions fit into the core of the three-dimensional structure.

Family (Structural Context)

As used in the FSSP database (Families of structurally similar proteins) and the DALI/ FSSP Web site, two structures that have a significant level of structural similarity but not necessarily significant sequence similarity.

Fold

Similar to structural motif, includes a larger combination of secondary structural units in the same configuration. Thus, proteins sharing the same fold have the same combination of secondary structures that are connected by similar loops. An example is the Rossman fold comprising several alternating α helices and parallel β strands. In the SCOP, CATH, and FSSP databases, the known protein structures have been classified into hierarchical levels of structural complexity with the fold as a basic level of classification.

Homologous Domain (Sequence Context)

An extended sequence pattern, generally found by sequence alignment methods, that indicates a common evolutionary origin among the aligned sequences. A homology domain is generally longer than motifs. The domain may include all of a given protein sequence or only a portion of the sequence. Some domains are complex and made up of several smaller homology domains that became joined to form a larger one during evolution. A domain that covers an entire sequence is called the homeomorphic domain by PIR (Protein Information Resource).

Module

A region of conserved amino acid patterns comprising one or more motifs and considered to be a fundamental unit of structure or function. The presence of a module has also been used to classify proteins into families.

Motif (Sequence Context)

A conserved pattern of amino acids that is found in two or more proteins. In the Prosite catalog, a motif is an amino acid pattern that is found in a group of proteins that have a similar biochemical activity, and that often is near the active site of the protein. Examples of sequence motif databases are the Prosite catalog and the Stanford Motifs Database.

Motif (Structural Context)

A combination of several secondary structural elements produced by the folding of adjacent sections of the polypeptide chain into a specific three-dimensional configuration. An example is the helix-loop-helix motif. Structural motifs are also referred to as supersecondary structures and folds.

Position-Specific Scoring Matrix (Sequence Context, Also Known As Weight or Scoring Matrix)

Represents a conserved region in a multiple sequence alignment with no gaps. Each matrix column represents the variation found in one column of the multiple sequence alignment.

Position-specific scoring matrix—3D (structural context) represents the amino acid variation found in an alignment of proteins that fall into the same structural class. Matrix columns represent the amino acid variation found at one amino acid position in the aligned structures.

Primary Structure

The linear amino acid sequence of a protein, which chemically is a polypeptide chain composed of amino acids joined by peptide bonds.

Profile (Sequence Context)

A scoring matrix that represents a multiple sequence alignment of a protein family. The profile is usually obtained from a well-conserved region in a multiple sequence alignment. The profile is in the form of a matrix with each column representing a position in the alignment and each row one of the amino acids. Matrix values give the likelihood of each amino acid at the corresponding position in the alignment. The profile is moved along the target sequence to locate the best scoring regions by a dynamic programming algorithm. Gaps are allowed during matching and a gap penalty is included in this case as a negative score when no amino acid is matched. A sequence profile may also be represented by a hidden Markov model, referred to as a profile HMM.

Profile (Structural Context)

A scoring matrix that represents which amino acids should fit well and which should fit poorly at sequential positions in a known protein structure. Profile columns represent sequential positions in the structure, and profile rows represent the 20 amino acids. As with a sequence profile, the structural profile is moved along a target sequence to find the highest possible alignment score by a dynamic programming algorithm. Gaps may be included and receive a penalty. The resulting score provides an indication as to whether or not the target protein might adopt such a structure.

Quaternary Structure

The three-dimensional configuration of a protein molecule comprising several independent polypeptide chains.

Secondary Structure

The interactions that occur between the C, O, and NH groups on amino acids in a polypeptide chain to form α-helices, β-sheets, turns, loops, and other forms, and that facilitate the folding into a three-dimensional structure.

Superfamily

A group of protein families of the same or different lengths that are related by distant yet detectable sequence similarity. Members of a given superfamily thus have a common evolutionary origin. Originally, Dayhoff defined the cutoff for superfamily status as being the chance that the sequences are not related of 10 6, on the basis of an alignment score (Dayhoff et al. 1978). Proteins with few identities in an alignment of the sequences but with a convincingly common number of structural and functional features are placed in the same superfamily. At the level of three-dimensional structure, superfamily proteins will share common structural features such as a common fold, but there may also be differences in the number and arrangement of secondary structures. The PIR resource uses the term homeomorphic superfamilies to refer to superfamilies that are composed of sequences that can be aligned from end to end, representing a sharing of single sequence homology domain, a region of similarity that extends throughout the alignment. This domain may also comprise smaller homology domains that are shared with other protein families and superfamilies. Although a given protein sequence may contain domains found in several superfamilies, thus indicating a complex evolutionary history, sequences will be assigned to only one homeomorphic superfamily based on the presence of similarity throughout a multiple sequence alignment. The superfamily alignment may also include regions that do not align either within or at the ends of the alignment. In contrast, sequences in the same family align well throughout the alignment.

Supersecondary Structure

A term with similar meaning to a structural motif. Tertiary structure is the three-dimensional or globular structure formed by the packing together or folding of secondary structures of a polypeptide chain.

SECONDARY STRUCTURE

Secondary structure prediction is a set of techniques in bioinformatics that aim to predict the local secondary structures of proteins based only on knowledge of their amino acid sequence. For proteins, a prediction consists of assigning regions of the amino acid sequence as likely alpha helices, beta strands (often noted as "extended" conformations), or turns. The success of a prediction is determined by comparing it to the results of the DSSP algorithm (or similar e.g. STRIDE) applied to the crystal structure of the protein. Specialized algorithms have been developed for the detection of specific well-defined patterns such as transmembrane helices and coiled coils in proteins.

The best modern methods of secondary structure prediction in proteins reach about 80% accuracy; this high accuracy allows the use of the predictions as feature improving fold recognition and ab initio protein structure prediction, classification of structural motifs, and refinement of sequence alignments. The accuracy of current protein secondary structure prediction methods is assessed in weekly benchmarks such as LiveBench and EVA.

Early methods of secondary structure prediction, introduced in the 1960s and early 1970s, focused on identifying likely alpha helices and were based mainly on helix-coil transition models. Significantly more accurate predictions that included beta sheets were introduced in the 1970s and relied on statistical assessments based on probability parameters derived from known solved structures. These methods, applied to a single sequence, are typically at most about 60-65% accurate, and often underpredict beta sheets. The evolutionary conservation of secondary structures can be exploited by simultaneously assessing many homologous sequences in a multiple sequence alignment, by calculating the net secondary structure propensity of an aligned column of amino acids. In concert with larger databases of known protein structures and modern machine learning methods such as neural nets and support vector machines, these methods can achieve up 80% overall accuracy in globular proteins. The theoretical upper limit of accuracy is around 90%, partly due to idiosyncrasies in DSSP assignment near the ends of secondary structures, where local conformations vary under native conditions but may be forced to assume a single conformation in crystals due to packing constraints. Limitations are also imposed by secondary structure prediction's inability to account for tertiary structure; for example, a sequence predicted as a likely helix may still be able to adopt a beta-strand conformation if it is located within a beta-sheet region of the protein and its side chains pack well with their neighbors. Dramatic conformational changes related to the protein's function or environment can also alter local secondary structure.

Historical Perspective

To date, over 20 different secondary structure prediction methods have been developed. One of the first algorithms was Chou-Fasman method, which relies predominantly on probability parameters determined from relative frequencies of each amino acid's appearance in each type of secondary structure. The original Chou-Fasman parameters, determined from the small sample of structures solved in the mid-1970s, produce poor results compared to modern methods, though the parameterization has been updated since it was first published. The Chou-Fasman method is roughly 50-60% accurate in predicting secondary structures.

The next notable program was the GOR method, named for the three scientists who developed it — Garnier, Osguthorpe, and Robson, is an information theory-based method. It uses the more powerful probabilistic technique of Bayesian inference. The GOR method takes into account not only the probability of each amino acid having a particular secondary structure, but also the conditional probability of the amino acid assuming each structure given the contributions of its neighbors (it does not assume that the neighbors have that same structure). The approach is both more sensitive and more accurate than that of Chou and Fasman because amino acid structural propensities are only strong for a small number of amino acids such as proline and glycine. Weak contributions from each of many neighbors can add up to strong effects overall. The original GOR method was roughly 65% accurate and is dramatically more successful in predicting alpha helices than beta sheets, which it frequently mispredicted as loops or disorganized regions.

Another big step forward, was using machine learning methods. First artificial neural networks methods were used. As a training sets they use solved structures to identify common sequence motifs associated with particular arrangements of secondary structures. These methods are over 70% accurate in their predictions, although beta strands are still often underpredicted due to the lack of three-dimensional structural information that would allow assessment of hydrogen bonding patterns that can promote formation of the extended conformation required for the presence of a complete beta sheet. PSIPRED and JPRED are some of the most known programs based on neural networks for protein secondary structure prediction. Next, support vector machines have proven particularly useful for predicting the locations of turns, which are difficult to identify with statistical methods.

Extensions of machine learning techniques attempt to predict more fine-grained local properties of proteins, such as backbone dihedral angles in unassigned regions. Both SVMs and neural networks have been applied to this problem. More recently, real-value torsion angles can be accurately predicted by SPINE-X and successfully employed for ab initio structure prediction.

Other Improvements

It is reported that in addition to the protein sequence, secondary structure formation depends on other factors. For example, it is reported that secondary structure tendencies depend also on local environment, solvent accessibility of residues, protein structural class, and even the organism from which the proteins are obtained. Based on such observations, some studies have shown that secondary structure prediction can be improved by addition of information about protein structural class, residue accessible surface area and also contact number information.

TERTIARY STRUCTURE

The practical role of protein structure prediction is now more important than ever. Massive amounts of protein sequence data are produced by modern large-scale DNA sequencing efforts such as the Human Genome Project. Despite community-wide efforts in structural genomics, the output of experimentally determined protein structures—typically by time-consuming and relatively expensive X-ray crystallography or NMR spectroscopy—is lagging far behind the output of protein sequences.

The protein structure prediction remains an extremely difficult and unresolved undertaking. The two main problems are calculation of protein free energy and finding the global minimum of this energy. A protein structure prediction method must explore the space of possible protein structures which is astronomically large. These problems can be partially bypassed in "comparative" or homology modeling and fold recognition methods, in which the search space is pruned by the assumption that the protein in question adopts a structure that is close to the experimentally determined structure of another homologous protein. On the other hand, the de novo or ab initio protein structure prediction methods must explicitly resolve these problems. The progress and challenges in protein structure prediction has been reviewed in Zhang 2008.

Energy- and Fragment-Based Methods

Ab initio- or de novo- protein modelling methods seek to build three-dimensional protein models "from scratch", i.e., based on physical principles rather than (directly) on previously solved structures. There are many possible procedures that either attempt to mimic protein folding or apply some stochastic method to search possible solutions (i.e., global optimization of a suitable energy function). These procedures tend to require vast computational resources, and have thus only been carried out for tiny proteins. To predict protein structure de novo for larger proteins will require better algorithms and larger computational resources like those afforded by either powerful supercomputers (such as Blue Gene or MDGRAPE-3) or distributed computing (such as Folding@home, the Human Proteome Folding Project and Rosetta@Home). Although these computational barriers are vast, the potential benefits of structural genomics (by predicted or experimental methods) make ab initio structure prediction an active research field.

As of 2009, a 50-residue protein could be simulated atom-by-atom on a supercomputer for 1 millisecond. As of 2012, comparable stable-state sampling could be done on a standard desktop with a new graphics card and more sophisticated algorithms. A much larger simulation timescales can be achieved using coarse-grained modeling.

Evolutionary Covariation to Predict 3D Contacts

As sequencing became more commonplace in the 1990s several groups used protein sequence alignments to predict correlated mutations and it was hoped that these coevolved residues could be used to predict tertiary structure (using the analogy to distance constraints from experimental procedures such as NMR). The assumption is when single residue mutations are slightly deleterious, compensatory mutations may occur to restabilize residue-residue interactions. This early work used what are known as local methods to calculate correlated mutations from protein sequences, but suffered from indirect false correlations which result from treating each pair of residues as independent of all other pairs.

In 2011, a different, and this time global statistical approach, demonstrated that predicted coevolved residues were sufficient to predict the 3D fold of a protein, providing there are enough sequences available (>1,000 homologous sequences are needed). The method, EVfold, uses no homology modeling, threading or 3D structure fragments and can be run on a standard personal computer even for proteins with hundreds of residues. The accuracy of the contacts predicted using this and related approaches has now been demonstrated on many known structures and contact maps, including the prediction of experimentally unsolved transmembrane proteins.

Comparative Protein Modeling

Comparative protein modelling uses previously solved structures as starting points, or templates. This is effective because it appears that although the number of actual proteins is vast, there is a limited set of tertiarystructural motifs to which most proteins belong. It has been suggested that there are only around 2,000 distinct protein folds in nature, though there are many millions of different proteins.

These methods may also be split into two groups:

Homology Modeling

It is based on the reasonable assumption that two homologous proteins will share very similar structures. Because a protein's fold is more evolutionarily conserved than its amino acid sequence, a target sequence can be modeled with reasonable accuracy on a very distantly related template, provided that the relationship between target and template can be discerned through sequence alignment. It has been suggested that the primary bottleneck in comparative modelling arises from difficulties in alignment rather than from errors in structure prediction given a known-good alignment. Unsurprisingly, homology modelling is most accurate when the target and template have similar sequences.

Protein Threading

It scans the amino acid sequence of an unknown structure against a database of solved structures. In each case, a scoring function is used to assess the compatibility of the sequence to the structure, thus yielding possible three-dimensional models. This type of method is also known as 3D-1D fold recognition due to its compatibility analysis between three-dimensional structures and linear protein sequences. This method has also given rise to methods performing an inverse folding search by evaluating the compatibility of a given structure with a large database of sequences, thus predicting which sequences have the potential to produce a given fold.

Side-Chain Geometry Prediction

Accurate packing of the amino acid side chains represents a separate problem in protein structure prediction. Methods that specifically address the problem of predicting side-chain geometry include dead-end elimination and the self-consistent mean field methods. The side chain conformations with low energy are usually determined on the rigid polypeptide backbone and using a set of discrete side chain conformations known as "rotamers." The methods attempt to identify the set of rotamers that minimize the model's overall energy.

These methods use rotamer libraries, which are collections of favorable conformations for each residue type in proteins. Rotamer libraries may contain information about the conformation, its frequency, and the standard deviations about mean dihedral angles, which can be used in sampling. Rotamer libraries are derived from structural bioinformatics or other statistical analysis of side-chain conformations in known experimental structures of proteins, such as by clustering the observed conformations for tetrahedral carbons near the staggered (60°, 180°, -60°) values.

Rotamer libraries can be backbone-independent, secondary-structure-dependent, or backbone-dependent. Backbone-independent rotamer libraries make no reference to backbone conformation, and are calculated from all available side chains of a certain type (for instance, the first example of a rotamer library, done by Ponder and Richards at Yale in 1987). Secondary-structure-dependent libraries present different dihedral angles and/or rotamer frequencies for α-helix, β-sheet, or coil secondary structures. Backbone-dependent rotamer libraries present conformations and/or frequencies dependent on the local backbone conformation as defined by the backbone dihedral angles ϕ and ψ, regardless of secondary structure.

The modern versions of these libraries as used in most software are presented as multidimensional distributions of probability or frequency, where the peaks correspond to the dihedral-angle conformations considered as individual rotamers in the lists. Some versions are based on very carefully curated data and are used primarily for structure validation, while others emphasize relative frequencies in much larger data sets and are the form used primarily for structure prediction, such as the Dunbrack rotamer libraries.

Side-chain packing methods are most useful for analyzing the protein's hydrophobic core, where side chains are more closely packed; they have more difficulty addressing the looser constraints and higher flexibility of surface residues, which often occupy multiple rotamer conformations rather than just one.

Prediction of Structural Classes

Statistical methods have been developed for predicting structural classes of proteins based on their amino acid composition, pseudo amino acid composition and functional domain composition.

QUATERNARY STRUCTURE

In the case of complexes of two or more proteins, where the structures of the proteins are known or can be predicted with high accuracy, protein–protein docking methods can be used to predict the structure of the complex. Information of the effect of mutations at specific sites on the affinity of the complex helps to understand the complex structure and to guide docking methods.

Software

A great number of software tools for protein structure prediction exist. Approaches include homology modeling, protein threading, ab initio methods, secondary structure prediction, and transmembrane helix and signal peptide prediction. Two most successful methods based on CASP experiment are I-TASSER and HHpred.

Evaluation of Automatic Structure Prediction Servers

CASP, which stands for Critical Assessment of Techniques for Protein Structure Prediction, is a community-wide experiment for protein structure prediction taking place every two years since 1994. CASP provides with an opportunity to assess the quality of available human, non-automated methodology (human category) and automatic servers for protein structure prediction (server category, introduced in the CASP7). The official results of automated assessment in 2012 CASP10 are available at for automated servers and for human and server predictors. In December 2014 next CASP11 assessment will be publicly available.

The CAMEO3D Continuous Automated Model EvaluatiOn Server evaluates automated protein structure prediction servers on a weekly basis using blind predictions for newly release protein structures. CAMEO publishes the results on its website ().

STRUCTURAL ALIGNMENT

Structural alignment attempts to establish homology between two or more polymer structures based on their shape and three-dimensional conformation. This process is usually applied to protein tertiary structures but can also be used for large RNA molecules. In contrast to simple structural superposition, where at least some equivalent residues of the two structures are known, structural alignment requires no a priori knowledge of equivalent positions. Structural alignment is a valuable tool for the comparison of proteins with low sequence similarity, where evolutionary relationships between proteins cannot be easily detected by standard sequence alignment techniques. Structural alignment can therefore be used to imply evolutionary relationships between proteins that share very little common sequence. However, caution should be used in using the results as evidence for shared evolutionary ancestry because of the possible confounding effects of convergent evolution by which multiple unrelated amino acid sequences converge on a common tertiary structure.

Structural alignments can compare two sequences or multiple sequences. Because these alignments rely on information about all the query sequences' three-dimensional conformations, the method can only be used on sequences where these structures are known. These are usually found by X-ray crystallography or NMR spectroscopy. It is possible to perform a structural alignment on structures produced by structure prediction methods. Indeed, evaluating such predictions often requires a structural alignment between the model and the true known structure to assess the model's quality. Structural alignments are especially useful in analyzing data from structural genomics and proteomics efforts, and they can be used as comparison points to evaluate alignments produced by purely sequence-based bioinformatics methods.

The outputs of a structural alignment are a superposition of the atomic coordinate sets and a minimal root mean square deviation (RMSD) between the structures. The RMSD of two aligned structures indicates their divergence from one another. Structural alignment can be complicated by the existence of multiple protein domains within one or more of the input structures, because changes in relative orientation of the domains between two structures to be aligned can artificially inflate the RMSD.

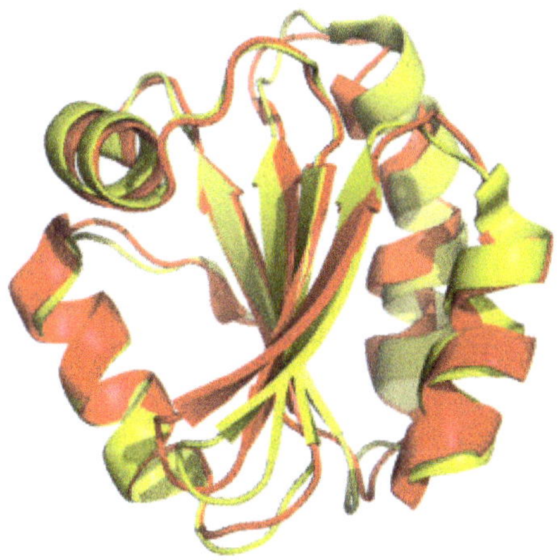

Structural alignment of thioredoxins from humans and the fly Drosophila melanogaster. The proteins are shown as ribbons, with the human protein in red, and the fly protein in yellow. Generated from PDB 3TRX and 1XWC.

Data Produced by Structural Alignment

The minimum information produced from a successful structural alignment is a set of superposed three-dimensional coordinates for each input structure. (Note that one input element may be fixed as a reference and therefore its superposed coordinates do not change.) The fitted structures can be used to calculate mutual RMSD values, as well as other more sophisticated measures of structural similarity such as the global distance test (GDT, the metric used in CASP). The structural alignment also implies a corresponding one-dimensional sequence alignment from which a sequence identity, or the percentage of residues that are identical between the input structures, can be calculated as a measure of how closely the two sequences are related.

Types of Comparisons

Because protein structures are composed of amino acids whose side chains are linked by a common protein backbone, a number of different possible subsets of the atoms that make up a protein macromolecule can be used in producing a structural alignment and calculating the corresponding RMSD values. When aligning structures with very different sequences, the side chain atoms generally are not taken into account because their identities differ between many aligned residues. For this reason it is common for structural alignment methods to use by default only the backbone atoms included in the peptide bond. For simplicity and efficiency, often only the alpha carbon positions are considered, since the peptide bond has a minimally variant planar conformation. Only when the structures to be aligned are highly similar or even identical is it meaningful to align side-chain atom positions, in which case the RMSD reflects not only the conformation of the protein backbone but also the rotameric states of the side chains. Other comparison criteria that reduce noise and bolster positive matches include secondary structure assignment, native contact maps or residue interaction patterns, measures of side chain packing, and measures of hydrogen bond retention.

Structural Superposition

The most basic possible comparison between protein structures makes no attempt to align the input structures and requires a precalculated alignment as input to determine which of the residues in the sequence are intended to be considered in the RMSD calculation. Structural superposition is commonly used to compare multiple conformations of the same protein (in which case no alignment is necessary, since the sequences are the same) and to evaluate the quality of alignments produced using only sequence information between two or more sequences whose structures are known. This method traditionally uses a simple least-squares fitting algorithm, in which the optimal rotations and translations are found by minimizing the sum of the squared distances among all structures in the superposition. More recently, maximum likelihood and Bayesian methods have greatly increased the accuracy of the estimated rotations, translations, and covariance matrices for the superposition.

Algorithms based on multidimensional rotations and modified quaternions have been developed to identify topological relationships between protein structures without the need for a predetermined alignment. Such algorithms have successfully identified canonical folds such as the four-helix bundle. The SuperPose method is sufficiently extensible to correct for relative domain rotations and other structural pitfalls.

Algorithmic Complexity

Optimal Solution

The optimal "threading" of a protein sequence onto a known structure and the production of an optimal multiple sequence alignment have been shown to be NP-complete. However, this does not imply that the structural alignment problem is NP-complete. Strictly speaking, an optimal solution to the protein structure alignment problem is only known for certain protein structure similarity measures, such as the measures used in protein structure prediction experiments, GDT_TS and MaxSub. These measures can be rigorously optimized using an algorithm capable of maximizing the number of atoms in two proteins that can be superimposed under a predefined distance cutoff. Unfortunately, the algorithm for optimal solution is not practical, since its running time depends not only on the lengths but also on the intrinsic geometry of input proteins.

Approximate Solution

Approximate polynomial-time algorithms for structural alignment that produce a family of "optimal" solutions within an approximation parameter for a given scoring function have been developed. Although these algorithms theoretically classify the approximate protein structure alignment problem as "tractable", they are still computationally too expensive for large-scale protein structure analysis. As a consequence, practical algorithms that converge to the global solutions of the alignment, given a scoring function, do not exist. Most algorithms are, therefore, heuristic, but algorithms that guarantee the convergence to at least local maximizers of the scoring functions, and are practical, have been developed.

Representation of Structures

Protein structures have to be represented in some coordinate-independent space to make them comparable. This is typically achieved by constructing a sequence-to-sequence matrix or series of matrices that encompass comparative metrics: rather than absolute distances relative to a fixed coordinate space. An intuitive representation is the distance matrix, which is a two-dimensional matrix containing all pairwise distances between some subset of the atoms in each structure (such as the alpha carbons). The matrix increases in dimensionality as the number of structures to be simultaneously aligned increases. Reducing the protein to a coarse metric such as secondary structure elements (SSEs) or structural fragments can also produce sensible alignments, despite the loss of information from discarding distances, as noise is also discarded. Choosing a representation to facilitate computation is critical to developing an efficient alignment mechanism.

Methods

Structural alignment techniques have been used in comparing individual structures or sets of structures as well as in the production of "all-to-all" comparison databases that measure the divergence between every pair of structures present in the Protein Data Bank (PDB). Such databases are used to classify proteins by their fold.

DALI

A common and popular structural alignment method is the DALI, or distance alignment matrix method, which breaks the input structures into hexapeptide fragments and calculates a distance matrix by evaluating the contact patterns between successive fragments. Secondary structure features that involve residues that are contiguous in sequence appear on the matrix's main diagonal; other diagonals in the matrix reflect spatial contacts between residues that are not near each other in the sequence. When these diagonals are parallel to the main diagonal, the features they represent are parallel; when they are perpendicular, their features are antiparallel. This representation is memory-intensive because the features in the square matrix are symmetrical (and thus redundant) about the main diagonal.

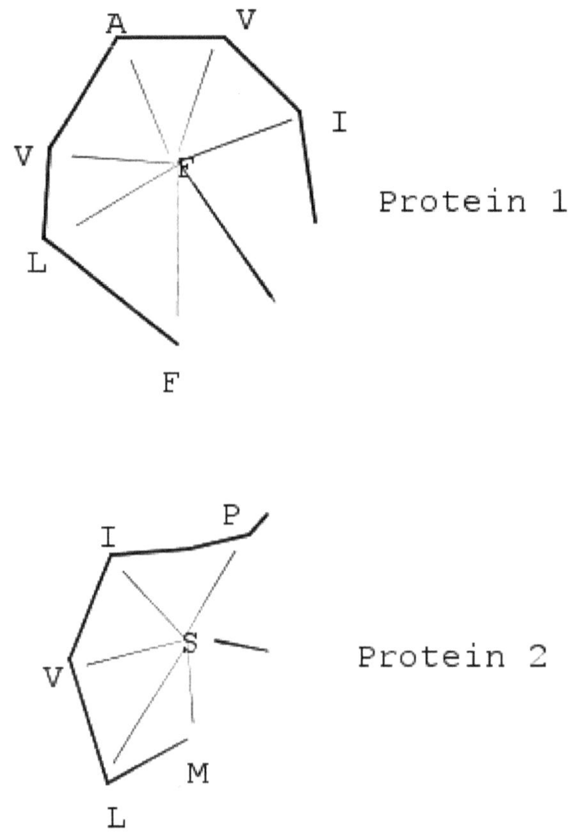

Illustration of the atom-to-atom vectors calculated in SSAP. From these vectors a series of vector differences, e.g., between (FA) in Protein 1 and (SI) in Protein 2 would be constructed. The two sequences are plotted on the two dimensions of a matrix to form a difference matrix between the two proteins. Dynamic programming is applied to all possible difference matrices to construct a series of optimal local alignment paths that are then summed to form the summary matrix, on which a second round of dynamic programming is performed.

When two proteins' distance matrices share the same or similar features in approximately the same positions, they can be said to have similar folds with similar-length loops connecting their secondary structure elements. DALI's actual alignment process requires a similarity search after the two proteins' distance matrices are built; this is normally conducted via a series of overlapping submatrices of size 6x6. Submatrix matches are then reassembled into a final alignment via a standard score-maximization algorithm — the original version of DALI used a Monte Carlo simulation to maximize a structural similarity score that is a function of the distances between putative corresponding atoms. In particular, more distant atoms within corresponding features are exponentially downweighted to reduce the effects of noise introduced by loop mobility, helix torsions, and other minor structural variations. Because DALI relies on an all-to-all distance matrix, it can account for the possibility that structurally aligned features might appear in different orders within the two sequences being compared.

The DALI method has also been used to construct a database known as FSSP (Fold classification based on Structure-Structure alignment of Proteins, or Families of Structurally Similar Proteins) in which all known protein structures are aligned with each other to determine their structural neighbors and fold classification. There is an searchable database based on DALI as well as a downloadable program and web search based on a standalone version known as DaliLite.

Combinatorial Extension

The combinatorial extension (CE) method is similar to DALI in that it too breaks each structure in the query set into a series of fragments that it then attempts to reassemble into a complete alignment. A series of pairwise combinations of fragments called aligned fragment pairs, or AFPs, are used to define a similarity matrix through which an optimal path is generated to identify the final alignment. Only AFPs that meet given criteria for local similarity are included in the matrix as a means of reducing the necessary search space and thereby increasing efficiency. A number of similarity metrics are possible; the original definition of the CE method included only structural superpositions and inter-residue distances but has since been expanded to include local environmental properties such as secondary structure, solvent exposure, hydrogen-bonding patterns, and dihedral angles.

An alignment path is calculated as the optimal path through the similarity matrix by linearly progressing through the sequences and extending the alignment with the next possible high-scoring AFP pair. The initial AFP pair that nucleates the alignment can occur at any point in the sequence matrix. Extensions then proceed with the next AFP that meets given distance criteria restricting the alignment to low gap sizes. The size of each AFP and the maximum gap size are required input parameters but are usually set to empirically determined values of 8 and 30 respectively. Like DALI and SSAP, CE has been used to construct an all-to-all fold classification database from the known protein structures in the PDB.

The RCSB PDB has recently released an updated version of CE and FATCAT as part of the RCSB PDB Protein Comparison Tool. It provides a new variation of CE that can detect circular permutations in protein structures.

SSAP

The SSAP (Sequential Structure Alignment Program) method uses double dynamic programming to produce a structural alignment based on atom-to-atom vectors in structure space. Instead of the alpha carbons typically used in structural alignment, SSAP constructs its vectors from the beta carbons for all residues except glycine, a method which thus takes into account the rotameric state of each residue as well as its location along the backbone. SSAP works by first constructing a series of inter-residue distance vectors between each residue and its nearest non-contiguous neighbors on each protein. A series of matrices are then constructed containing the vector differences between neighbors for each pair of residues for which vectors were constructed. Dynamic programming applied to each resulting matrix determines a series of optimal local alignments which are then summed into a "summary" matrix to which dynamic programming is applied again to determine the overall structural alignment.

SSAP originally produced only pairwise alignments but has since been extended to multiple alignments as well. It has been applied in an all-to-all fashion to produce a hierarchical fold classification scheme known as CATH (Class, Architecture, Topology, Homology), which has been used to construct the CATH Protein Structure Classification database.

Recent Developments

Improvements in structural alignment methods constitute an active area of research, and new or modified methods are often proposed that are claimed to offer advantages over the older and more widely distributed techniques. A recent example, TM-align, uses a novel method for weighting its distance matrix, to which standard dynamic programming is then applied. The weighting is proposed to accelerate the convergence of dynamic programming and correct for effects arising from alignment lengths. In a benchmarking study, TM-align has been reported to improve in both speed and accuracy over DALI and CE.

Other promising methods of structural alignment are local structural alignment methods. These provide comparison of pre-selected parts of proteins (e.g. binding sites, user-defined structural motifs) against binding sites or whole-protein structural databases. The MultiBind and MAPPIS servers allow the identification of common spatial arrangements of physicochemical properties such as H-bond donor, acceptor, aliphatic, aromatic or hydrophobic in a set of user provided protein binding sites defined by interactions with small molecules (MultiBind) or in a set of user-provided protein–protein interfaces (MAPPIS). Others provide comparison of entire protein structures against a number of user submitted structures or against a large database of protein structures in reasonable time (ProBiS). Unlike global alignment approaches, local structural alignment approaches are suited to detection of locally conserved patterns of functional groups, which often appear in binding sites and have significant involvement in ligand binding. As an example, comparing G-Losa, a local structure alignment tool, with TM-align, a global structure alignment based method. While G-Losa predicts drug-like ligands' positions in single-chain protein targets more precisely than TM-align, the overall success rate of TM-align is better.

However, as algorithmic improvements and computer performance have erased purely technical deficiencies in older approaches, it has become clear that there is no one universal criterion for the 'optimal' structural alignment. TM-align, for instance, is particularly robust in quantifying comparisons between sets of proteins with great disparities in sequence lengths, but it only indirectly captures hydrogen bonding or secondary structure order conservation which might be better metrics for alignment of evolutionarily related proteins. Thus recent developments have focused on optimizing particular attributes such as speed, quantification of scores, correlation to alternative gold standards, or tolerance of imperfection in structural data or ab initio structural models. An alternative methodology that is gaining popularity is to use the consensus of various methods to ascertain proteins structural similarities.

RNA Structural Alignment

Structural alignment techniques have traditionally been applied exclusively to proteins, as the primary biological macromolecules that assume characteristic three-dimensional structures. However, large RNA molecules also form characteristic tertiary structures, which are mediated primarily by hydrogen bonds formed between base pairs as well as base stacking. Functionally similar noncoding RNA molecules can be especially difficult to extract from genomics data because structure is more strongly conserved than sequence in RNA as well as in proteins, and the more limited alphabet of RNA decreases the information content of any given nucleotide at any given position.

However, because of the increasing interest in RNA structures and because of the growth of the number of experimentally determined 3D RNA structures, few RNA structure similarity methods have been developed recently. One of those methods is, e.g., SETTER which decomposes each RNA structure into smaller parts called general secondary structure units (GSSUs). GSSUs are subsequently aligned and these partial alignments are merged into the final RNA structure alignment and scored. The method has been implemented into the SETTER webserver.

A recent method for pairwise structural alignment of RNA sequences with low sequence identity has been published and implemented in the program FOLDALIGN. However, this method is not truly analogous to protein structural alignment techniques because it computationally predicts the structures of the RNA input sequences rather than requiring experimentally determined structures as input. Although computational prediction of the protein folding process has not been particularly successful to date, RNA structures without pseudoknots can often be sensibly predicted using free energy-based scoring methods that account for base pairing and stacking.

Software

Choosing a software tool for structural alignment can be a challenge due to the large variety of available packages that differ significantly in methodology and reliability. A partial solution to this problem was presented in and made publicly accessible through the ProCKSI webserver. A more complete list of currently available and freely distributed structural alignment software can be found in structural alignment software.

Properties of some structural alignment servers and software packages are summarized and tested with examples at Structural Alignment Tools in Proteopedia.Org.

MACROMOLECULAR DOCKING

Macromolecular docking is the computational modelling of the quaternary structure of complexes formed by two or more interacting biological macromolecules. Protein–protein complexes are the most commonly attempted targets of such modelling, followed by protein–nucleic acid complexes.

The ultimate goal of docking is the prediction of the three-dimensional structure of the macromolecular complex of interest as it would occur in a living organism. Docking itself only produces plausible candidate structures. These candidates must be ranked using methods such as scoring functions to identify structures that are most likely to occur in nature.

The term "docking" originated in the late 1970s, with a more restricted meaning; then, "docking" meant refining a model of a complex structure by optimizing the separation between the interactors but keeping their relative orientations fixed. Later, the relative orientations of the interacting partners in the modelling was allowed to vary, but the internal geometry of each of the partners was held fixed. This type of modelling is sometimes referred to as "rigid docking". With further increases in computational power, it became possible to model changes in internal geometry of the interacting partners that may occur when a complex is formed. This type of modelling is referred to as "flexible docking".

Background

The biological roles of most proteins, as characterized by which other macromolecules they interact with, are known at best incompletely. Even those proteins that participate in a well-studied biological process (e.g., the Krebs cycle) may have unexpected interaction partners or functions which are unrelated to that process.

In cases of known protein–protein interactions, other questions arise. Genetic diseases (e.g., cystic fibrosis) are known to be caused by misfolded or mutated proteins, and there is a desire to understand what, if any, anomalous protein–protein interactions a given mutation can cause. In the distant future, proteins may be designed to perform biological functions, and a determination of the potential interactions of such proteins will be essential.

For any given set of proteins, the following questions may be of interest, from the point of view of technology or natural history:

- Do these proteins bind *in vivo*?

If they do bind,

- What is the spatial configuration which they adopt in their bound state?
- How strong or weak is their interaction?

If they do not bind,

- Can they be made to bind by inducing a mutation?

Protein–protein docking is ultimately envisaged to address all these issues. Furthermore, since docking methods can be based on purely physical principles, even proteins of unknown function (or which have been studied relatively little) may be docked. The only prerequisite is that their molecular structure has been either determined experimentally, or can be estimated by a protein structure prediction technique.

Protein–nucleic acid interactions feature prominently in the living cell. Transcription factors, which regulate gene expression, and polymerases, which catalyse replication, are composed of proteins, and the genetic material they interact with is composed of nucleic acids. Modeling protein–nucleic acid complexes presents some unique challenges, as described below.

History

In the 1970s, complex modelling revolved around manually identifying features on the surfaces of the interactors, and interpreting the consequences for binding, function and activity; any computer programmes were typically used at the end of the modelling process, to discriminate between the relatively few configurations which remained after all the heuristic constraints had been imposed. The first use of computers was in a study on hemoglobin interaction in sickle-cell fibres. This was followed in 1978 by work on the trypsin-BPTI complex. Computers discriminated between good and bad models using a scoring function which rewarded large interface area, and pairs of molecules in contact but not occupying the same space. The computer used a simplified representation of the interacting proteins, with one interaction centre for each residue. Favorable electrostatic interactions, including hydrogen bonds, were identified by hand.

In the early 1990s, more structures of complexes were determined, and available computational power had increased substantially. With the emergence of bioinformatics, the focus moved towards developing generalized techniques which could be applied to an arbitrary set of complexes at acceptable computational cost. The new methods were envisaged to apply even in the absence of phylogenetic or experimental clues; any specific prior knowledge could still be introduced at the stage of choosing between the highest ranking output models, or be framed as input if the algorithm catered for it. 1992 saw the publication of the correlation method, an algorithm which used the fast Fourier transform to give a vastly improved scalability for evaluating coarse shape complementarity on rigid-body models. This was extended in 1997 to cover coarse electrostatics.

In 1996 the results of the first blind trial were published, in which six research groups attempted to predict the complexed structure of TEM-1 Beta-lactamase with Beta-lactamase inhibitor protein (BLIP). The exercise brought into focus the necessity of accommodating conformational change and the difficulty of discriminating between conformers. It also served as the prototype for the CAPRI assessment series, which debuted in 2001.

Rigid-Body Docking *vs.* Flexible Docking

If the bond angles, bond lengths and torsion angles of the components are not modified at any stage of complex generation, it is known as rigid body docking. A subject of speculation is whether or not rigid-body docking is sufficiently good for most docking. When substantial conformational change occurs within the components at the time of complex formation, rigid-body docking is inadequate. However, scoring all possible conformational changes is prohibitively expensive in computer time. Docking procedures which permit conformational change, or flexible docking procedures, must intelligently select small subset of possible conformational changes for consideration.

Methods

Successful docking requires two criteria:

- Generating a set of configurations which reliably includes at least one nearly correct one.
- Reliably distinguishing nearly correct configurations from the others.

For many interactions, the binding site is known on one or more of the proteins to be docked. This is the case for antibodies and for competitive inhibitors. In other cases, a binding site may be strongly suggested by mutagenic or phylogenetic evidence. Configurations where the proteins interpenetrate severely may also be ruled out a priori.

After making exclusions based on prior knowledge or stereochemical clash, the remaining space of possible complexed structures must be sampled exhaustively, evenly and with a sufficient coverage to guarantee a near hit. Each configuration must be scored with a measure that is capable of ranking a nearly correct structure above at least 100,000 alternatives. This is a computationally intensive task, and a variety of strategies have been developed.

Reciprocal Space Methods

Each of the proteins may be represented as a simple cubic lattice. Then, for the class of scores which are discrete convolutions, configurations related to each other by translation of one protein by an exact lattice vector can all be scored almost simultaneously by applying the convolution theorem. It is possible to construct reasonable, if approximate, convolution-like scoring functions representing both stereochemical and electrostatic fitness.

Reciprocal space methods have been used extensively for their ability to evaluate enormous numbers of configurations. They lose their speed advantage if torsional changes are introduced. Another drawback is that it is impossible to make efficient use of prior knowledge. The question also remains whether convolutions are too limited a class of scoring function to identify the best complex reliably.

Monte Carlo Methods

In Monte Carlo, an initial configuration is refined by taking random steps which are accepted or rejected based on their induced improvement in score, until a certain number of steps have been tried. The assumption is that convergence to the best structure should occur from a large class of initial configurations, only one of which needs to be considered. Initial configurations may be sampled coarsely, and much computation time can be saved. Because of the difficulty of finding a scoring function which is both highly discriminating for the correct configuration and also converges to the correct configuration from a distance, the use of two levels of refinement, with different scoring functions, has been proposed. Torsion can be introduced naturally to Monte Carlo as an additional property of each random move.

Monte Carlo methods are not guaranteed to search exhaustively, so that the best configuration may be missed even using a scoring function which would in theory identify it. How severe a problem this is for docking has not been firmly established.

Evaluation

Scoring Functions

To find a score which forms a consistent basis for selecting the best configuration, studies are carried out on a standard benchmark (see below) of protein–protein interaction cases. Scoring functions are assessed on the rank they assign to the best structure (ideally the best structure should be ranked 1), and on their coverage (the proportion of the benchmark cases for which they achieve an acceptable result). Types of scores studied include:

- Heuristic scores based on residue contacts.
- Shape complementarity of molecular surfaces ("stereochemistry").
- Free energies, estimated using parameters from molecular mechanics force fields such as CHARMM or AMBER.
- Phylogenetic desirability of the interacting regions.
- Clustering coefficients.

It is usual to create hybrid scores by combining one or more categories above in a weighted sum whose weights are optimized on cases from the benchmark. To avoid bias, the benchmark cases used to optimize the weights must not overlap with the cases used to make the final test of the score.

The ultimate goal in protein–protein docking is to select the ideal ranking solution according to a scoring scheme that would also give an insight into the affinity of the complex. Such a development would drive in silico protein engineering, computer-aided drug design and/or high-throughput annotation of which proteins bind or not (annotation of interactome). Several scoring functions have been proposed for binding affinity / free energy prediction. However the correlation between experimentally determined binding affinities and the predictions of nine commonly used scoring functions have been found to be nearly orthogonal (R2 ~ 0). It was also observed that some components of the scoring algorithms may display better correlation to the experimental binding energies than the full score, suggesting that a significantly better performance might be obtained by combining the appropriate contributions from different scoring algorithms. Experimental methods for the determination of binding affinities are: surface plasmon resonance (SPR), Förster resonance energy transfer, radioligand-based techniques, isothermal titration calorimetry (ITC), microscale thermophoresis (MST) or spectroscopic measurements and other fluorescence techniques.

Benchmarks

A benchmark of 84 protein–protein interactions with known complexed structures has been developed for testing docking methods. The set is chosen to cover a wide range of interaction types, and to avoid repeated features, such as the profile of interactors' structural families according to the SCOP database. Benchmark elements are classified into three levels of difficulty (the most difficult containing the largest change in backbone conformation). The protein–protein docking benchmark contains examples of enzyme-inhibitor, antigen-antibody and homomultimeric complexes.

The latest version of protein-protein docking benchmark consists of 230 complexes. A protein-DNA docking benchmark consists of 47 test cases. A protein-RNA docking benchmark was curated as a dataset of 45 non-redundant test cases with complexes solved by X-ray crystallography only as well as an extended dataset of 71 test cases with structures derived from homology modelling as well. The protein-RNA benchmark has been updated to include more structures solved by X-ray crystallography and now it consists of 126 test cases.

A binding affinity benchmark has been based on the protein–protein docking benchmark. 81 protein–protein complexes with known experimental affinities are included; these complexes span over 11 orders of magnitude in terms of affinity. Each entry of the benchmark includes several biochemical parameters associated with the experimental data, along with the method used to determine the affinity. This benchmark was used to assess the extent to which scoring functions could also predict affinities of macromolecular complexes.

This Benchmark was post-peer reviewed and significantly expanded. The new set is diverse in terms of the biological functions it represents, with complexes that involve G-proteins and receptor extracellular domains, as well as antigen/antibody, enzyme/inhibitor, and enzyme/substrate complexes. It is also diverse in terms of the partners' affinity for each other, with Kd ranging between 10−5 and 10−14 M. Nine pairs of entries represent closely related complexes that have a similar structure, but a very different affinity, each pair comprising a cognate and a noncognate assembly. The unbound structures of the component proteins being available, conformation changes can be assessed. They are significant in most of the complexes, and large movements or disorder-to-order transitions are frequently observed. The set may be used to benchmark biophysical models aiming to relate affinity to structure in protein–protein interactions, taking into account the reactants and the conformation changes that accompany the association reaction, instead of just the final product.

The CAPRI Assessment

The Critical Assessment of PRediction of Interactions is an ongoing series of events in which researchers throughout the community try to dock the same proteins, as provided by the assessors. Rounds take place approximately every 6 months. Each round contains between one and six target protein–protein complexes whose structures have been recently determined experimentally. The coordinates and are held privately by the assessors, with the cooperation of the structural biologists who determined them. The assessment of submissions is double blind.

CAPRI attracts a high level of participation (37 groups participated worldwide in round seven) and a high level of interest from the biological community in general. Although CAPRI results are of little statistical significance owing to the small number of targets in each round, the role of CAPRI in stimulating discourse is significant. (The CASP assessment is a similar exercise in the field of protein structure prediction).

PROTEIN STRUCTURE

Protein structure prediction is the inference of the three-dimensional structure of a protein from its amino acid sequence—that is, the prediction of its folding and its secondary and tertiary structure from its primary structure. Structure prediction is fundamentally different from the inverse problem of protein design.

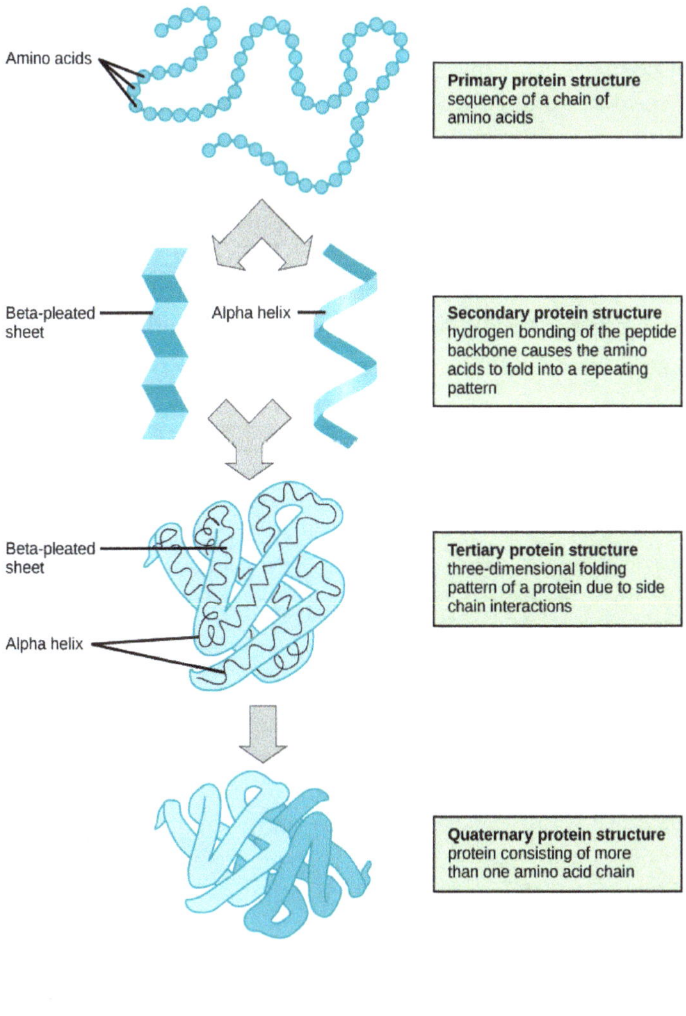

Constituent amino-acids can be analyzed to predict secondary, tertiary and quaternary protein structure.

Protein structure prediction is one of the most important goals pursued by bioinformatics and theoretical chemistry; it is highly important in medicine (for example, in drug design) and biotechnology (for example, in the design of novel enzymes). Every two years, the performance of current methods is assessed in the CASP experiment (Critical Assessment of Techniques for Protein Structure Prediction). A continuous evaluation of protein structure prediction web servers is performed by the community project CAMEO3D.

Protein Structure and Terminology

Proteins are chains of amino acids joined together by peptide bonds. Many conformations of this chain are possible due to the rotation of the chain about each Cα atom. It is these conformational changes that are responsible for differences in the three dimensional structure of proteins. Each amino acid in the chain is polar, i.e. it has separated positive and negative charged regions with a free C=O group, which can act as hydrogen bond acceptor and an NH group, which can act as hydrogen bond donor. These groups can therefore interact in the protein structure. The 20 amino acids can be classified according to the chemistry of the side chain which also plays an important structural role. Glycine takes on a special position, as it has the smallest side chain, only one Hydrogen atom, and therefore can increase the local flexibility in the protein structure. Cysteine on the other hand can react with another cysteine residue and thereby form a cross link stabilizing the whole structure.

The protein structure can be considered as a sequence of secondary structure elements, such as α helices and β sheets, which together constitute the overall three-dimensional configuration of the protein chain. In these secondary structures regular patterns of H bonds are formed between neighboring amino acids, and the amino acids have similar Φ and Ψ angles.

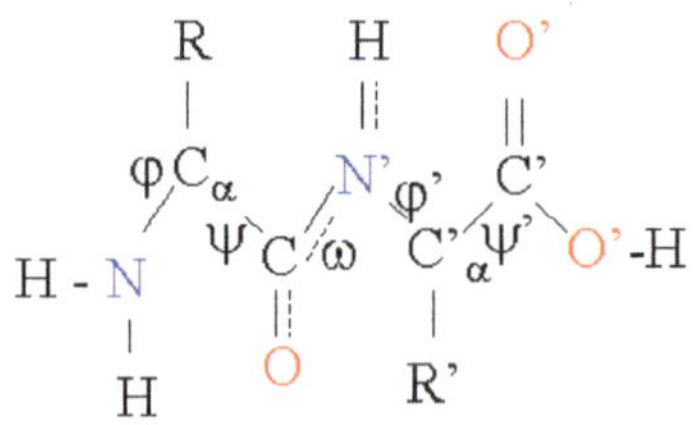

Bond angles for ψ and ω

The formation of these structures neutralizes the polar groups on each amino acid. The secondary structures are tightly packed in the protein core in a hydrophobic environment. Each amino acid side group has a limited volume to occupy and a limited number of possible interactions with other nearby side chains, a situation that must be taken into account in molecular modeling and alignments.

Alpha Helix

The alpha helix (α-helix) is a common motif in the secondary structure of proteins and is a righthand-coiled or spiral conformation (helix) in which every backbone N-H group donates a hydrogen bond to the backbone C=O group of the amino acid located three or four residues earlier along the protein sequence. This secondary structure is also sometimes called a classic Pauling–Corey–Branson α-helix (see below). The name 3.613-helix is also used for this type of helix, denoting the average number of residues per helical turn, with 13 atoms being involved in the ring formed by the hydrogen bond. Among types of local structure in proteins, the α-helix is the most regular and the most predictable from sequence, as well as the most prevalent.

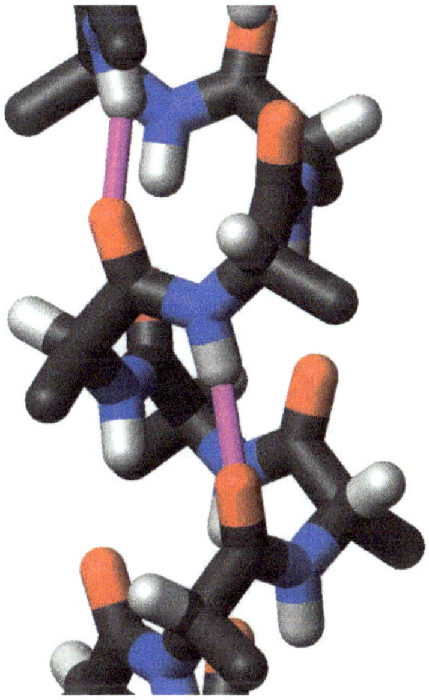

Side view of an **α-helix** of alanine residues in atomic detail. Two hydrogen bonds for the same peptide group are highlighted in magenta; the H to O distance is about 2 **Å** (0.20 nm). The protein chain runs upward here; that is, its N-terminus is at the bottom and its C-terminus at the top. Note that the sidechains (black stubs) angle slightly downward, toward the N-terminus, while the peptide oxygens (red) point up and the peptide NHs (blue with grey stubs) point down.

Discovery

In the early 1930s, William Astbury showed that there were drastic changes in the X-ray fiber diffraction of moist wool or hair fibers upon significant stretching. The data suggested that the unstretched fibers had a coiled molecular structure with a characteristic repeat of ~5.1 ångströms (0.51 nm).

Astbury initially proposed a kinked-chain structure for the fibers. He later joined other researchers (notably the American chemist Maurice Huggins) in proposing that:

- the unstretched protein molecules formed a helix (which he called the α-form)

- the stretching caused the helix to uncoil, forming an extended state (which he called the β-form).

Although incorrect in their details, Astbury's models of these forms were correct in essence and correspond to modern elements of secondary structure, the α-helix and the β-strand (Astbury's nomenclature was kept), which were developed by Linus Pauling, Robert Corey and Herman Branson in 1951 (see below); that paper showed both right- and left-handed helixes, although in 1960 the crystal structure of myoglobin showed that the right-handed form is the common one. Hans Neurath was the first to show that Astbury's models could not be correct in detail, because they involved clashes of atoms. Neurath's paper and Astbury's data inspired H. S. Taylor,Maurice Huggins and Bragg and collaborators to propose models of keratin that somewhat resemble the modern α-helix.

Two key developments in the modeling of the modern α-helix were (1) the correct bond geometry, thanks to the crystal structure determinations of amino acids and peptides and Pauling's prediction of planar peptide bonds; and (2) his relinquishing of the assumption of an integral number of residues per turn of the helix. The pivotal moment came in the early spring of 1948, when Pauling caught a cold and went to bed. Being bored, he drew a polypeptide chain of roughly correct dimensions on a strip of paper and folded it into a helix, being careful to maintain the planar peptide bonds. After a few attempts, he produced a model with physically plausible hydrogen bonds. Pauling then worked with Corey and Branson to confirm his model before publication. In 1954, Pauling was awarded his first Nobel Prize "for his research into the nature of the chemical bond and its application to the elucidation of the structure of complex substances" (such as proteins), prominently including the structure of the α-helix.

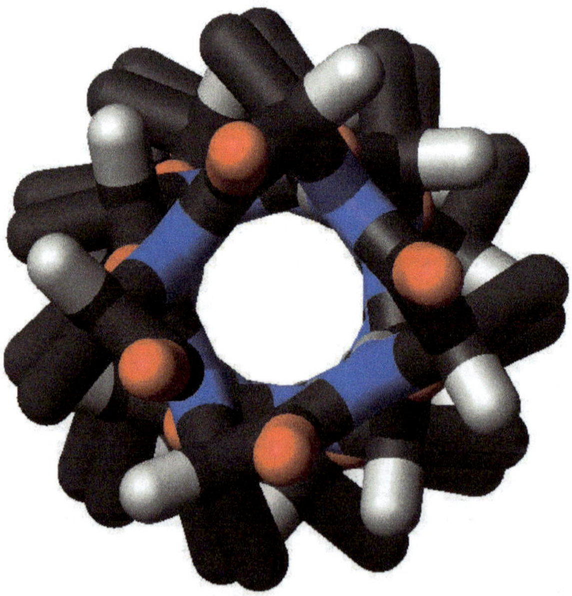

Top view of the same helix shown above. Four carbonyl groups are pointing upwards toward the viewer, spaced roughly 100° apart on the circle, corresponding to 3.6 amino-acid residues per turn of the helix.

Structure

Geometry and Hydrogen Bonding

The amino acids in an α-helix are arranged in a right-handed helical structure where each amino acid residue corresponds to a 100° turn in the helix (i.e., the helix has 3.6 residues per turn), and a translation of 1.5 Å (0.15 nm) along the helical axis. Dunitz describes how Pauling's first article on the theme in fact shows a left-handed helix, the enantiomer of the true structure. Short pieces of left-handed helix sometimes occur with a large content of achiral glycine amino acids, but are unfavorable for the other normal, biological L-amino acids. The pitch of the alpha-helix (the vertical distance between consecutive turns of the helix) is 5.4 Å (0.54 nm), which is the product of 1.5 and 3.6. What is most important is that the N-H group of an amino acid forms a hydrogen bond with the C=O group of the amino acid four residues earlier; this repeated $i+4 \rightarrow i$ hydrogen bonding is the most prominent characteristic of an α-helix. Official international nomenclature specifies two ways of defining α-helices, rule 6.2 in terms of repeating φ,ψ torsion angles (see below) and rule 6.3 in terms of the combined pattern of pitch and hydrogen bonding. The alpha-helices can be identified in protein structure using several computational methods, one of which being DSSP (Dictionary of Protein Secondary Structure).

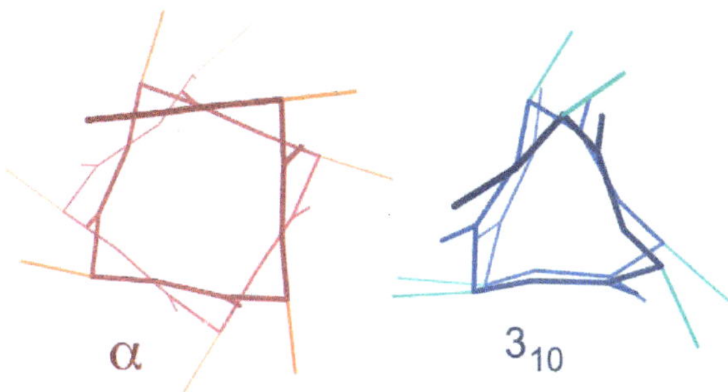

Contrast of helix end views between α (offset squarish) vs 3_{10} (triangular)

Similar structures include the 310 helix ($i + 3 \rightarrow i$ hydrogen bonding) and the π-helix ($i + 5 \rightarrow i$ hydrogen bonding). The α-helix can be described as a 3.613 helix, since the i + 4 spacing adds 3 more atoms to the H-bonded loop compared to the tighter 310helix, and on average, 3.6 amino acids are involved in one ring of α-helix. The subscripts refer to the number of atoms (including the hydrogen) in the closed loop formed by the hydrogen bond.

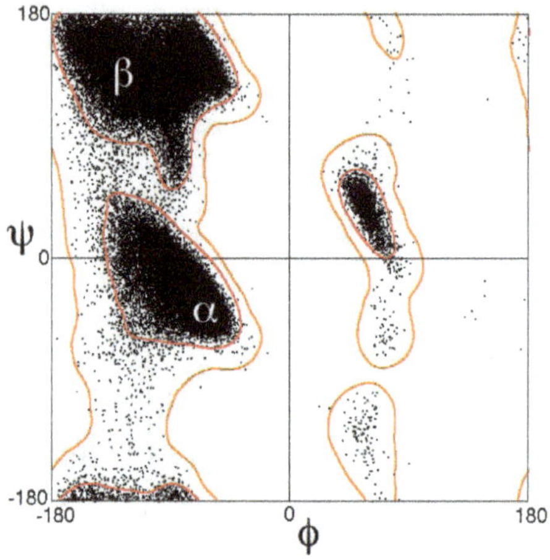

Ramachandran plot (φ,ψ plot), with data points for α-helical residues forming a dense diagonal cluster below and left of center, around the global energy minimum for backbone conformation.

Residues in α-helices typically adopt backbone (φ, ψ) dihedral angles around (-60°, -45°), as shown in the image at right. In more general terms, they adopt dihedral angles such that the ψ dihedral angle of one residue and the φ dihedral angle of the next residue sum to roughly -105°. As a consequence, α-helical dihedral angles, in general, fall on a diagonal stripe on the Ramachandran diagram (of slope -1), ranging from (-90°, -15°) to (-35°, -70°). For comparison, the sum of the dihedral angles for a 310 helix is roughly -75°, whereas that for the π-helix is roughly -130°. The general formula for the rotation angle Ω per residue of any polypeptide helix with trans isomers is given by the equation

$$3\cos\Omega = 1 - 4\cos^2[(\phi + \psi)/2]$$

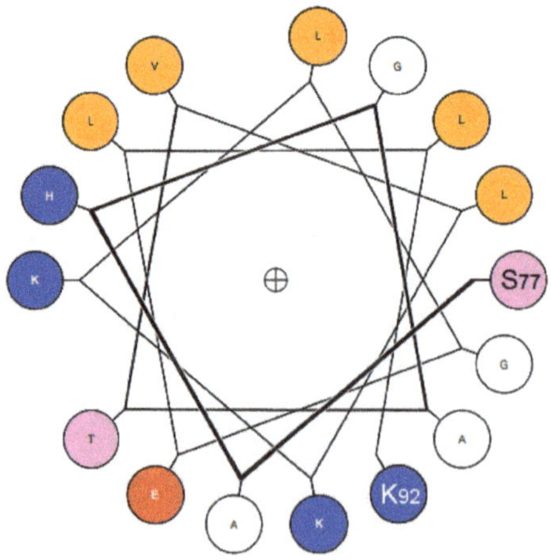

Helical wheel diagram

The α-helix is tightly packed; there is almost no free space within the helix. The amino-acid side-chains are on the outside of the helix, and point roughly "downward" (i.e., toward the N-terminus), like the branches of an evergreen tree (Christmas tree effect). This directionality is sometimes used in preliminary, low-resolution electron-density maps to determine the direction of the protein backbone.

2D (2-Dimensional) Diagrams for Representing α-Helices

Three differently arranged styles of 2D diagrams are used to represent different aspects of the sequence and structure relationships that confer specific physical and interaction properties on individual α-helices. Two of these emphasize circular placement around the cylindrical cross-section: The first-developed such diagram is called the "helical wheel", and a more recent version is called the "wenxiang diagram". The latter name came from the fact that it resembles a coil-like incense used in China to repel mosquitos.

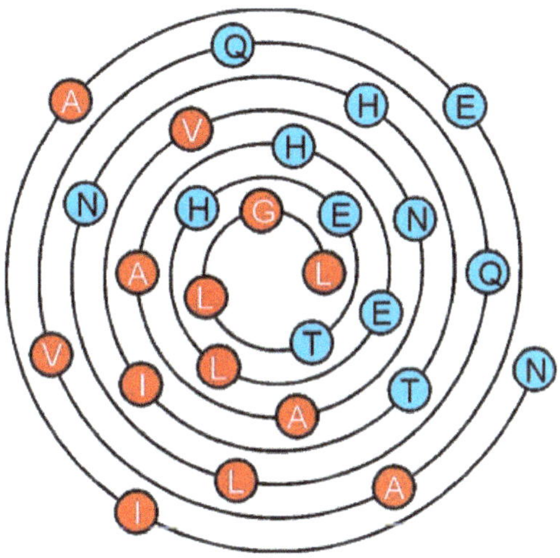

Wenxiang diagram

The helical wheel represents a helix by a projection of the Cα backbone structure down the helix axis, while the wenxiang diagram represents it more abstractly as a smooth spiral coiled on the plane of the page. Both label the sequence with one-letter amino-acid code (see amino acid) at each Cα position, using different colors or symbols to code the amino-acid properties. Hydrophobic vs hydrophilic amino acids are always distinguished, as the most important property governing helix interactions. Sometimes positively vs negatively charged hydrophilics are distinguished, and sometimes ambiguous amino acids such as glycine (G) are distinguished. Color-coding conventions are various. The helical wheel does not change representation along the helix, while the wenxiang diagram is able to show the relative locations of the amino acids in an α-helix regardless of how long it is.

Either circular style of diagram can provide an intuitive and easily visualizable 2D picture that characterizes the disposition of hydrophobic and hydrophilic residues in α-helices, and can be used to study helix-helix interactions, helix-membrane interactions as quantified by the helical hydrophobic moment, or protein-protein interactions. Various utilities and web sites are available to generate helical wheels, such as the page by Kael Fischer.

The third style of 2D diagram is called a "helical net". It is generated by opening the cylindrical surface of each helix along a line parallel to the axis and laying the result out vertically. The helix net is not suitable for studying helix-helix packing interactions, but it has become the dominant means of representing the sequence arrangement for integral membrane proteins because it shows important relationships of the helical sequence to vertical positioning within the membrane even without knowledge of how the helices are arranged in 3D.

Stability

Helices observed in proteins can range from four to over forty residues long, but a typical helix contains about ten amino acids (about three turns). In general, short polypeptides do not exhibit much α-helical structure in solution, since the entropic cost associated with the folding of the polypeptide chain is not compensated for by a sufficient amount of stabilizing interactions.

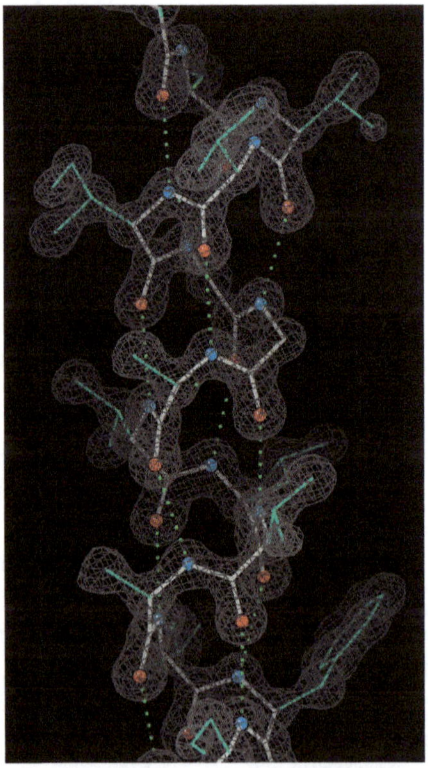

An α-helix in ultra-high-resolution electron density contours, with O atoms in red, N atoms in blue, and hydrogen bonds as green dotted lines (PDB file 2NRL, 17-32). The N-terminus is at the top, here.

In general, the backbone hydrogen bonds of α-helices are considered slightly weaker than those found in β-sheets, and are readily attacked by the ambient water molecules. However, in more hydrophobic environments such as the plasma membrane, or in the presence of co-solvents such as trifluoroethanol (TFE), or isolated from solvent in the gas phase, oligopeptides readily adopt stable α-helical structure. Furthermore, crosslinks can be incorporated into peptides to conformationally stabilize helical folds. Crosslinks stabilize the helical state by entropically destabilizing the unfolded state and by removing enthalpically stabilized "decoy" folds that compete with the fully helical state.

Experimental Determination

Since the α-helix is defined by its hydrogen bonds and backbone conformation, the most detailed experimental evidence for α-helical structure comes from atomic-resolution X-ray crystallography such as the example shown at right. It is clear that all the backbone carbonyl oxygens point downward (toward the C-terminus) but splay out slightly, and the H-bonds are approximately parallel to the helix axis. Protein structures from NMR spectroscopy also show helices well, with characteristic observations of nuclear Overhauser effect (NOE) couplings between atoms on adjacent helical turns. In some cases, the individual hydrogen bonds can be observed directly as a small scalar coupling in NMR.

There are several lower-resolution methods for assigning general helical structure. The NMR chemical shifts (in particular of the C^{α}, C^{β} and C' atoms) and residual dipolar couplings are often characteristic of helices. The far-UV (170-250 nm) circular dichroism spectrum of helices is also idiosyncratic, exhibiting a pronounced double minimum at ~208 nm and ~222 nm. Infrared spectroscopy is rarely used, since the α-helical spectrum resembles that of a random coil (although these might be discerned by, e.g., hydrogen-deuterium exchange). Finally, cryo electron microscopy is now capable of discerning individual α-helices within a protein, although their assignment to residues is still an active area of research.

Long homopolymers of amino acids often form helices if soluble. Such long, isolated helices can also be detected by other methods, such as dielectric relaxation, flow birefringence, and measurements of the diffusion constant. In stricter terms, these methods detect only the characteristic prolate (long cigar-like) hydrodynamic shape of a helix, or its large dipole moment.

Amino-Acid Propensities

Different amino-acid sequences have different propensities for forming α-helical structure. Methionine, alanine, leucine, glutamate, and lysine uncharged ("MALEK" in the amino-acid 1-letter codes) all have especially high helix-forming propensities, whereas proline and glycine have poor helix-forming propensities. Proline either breaks or kinks a helix, both because it cannot donate an amide hydrogen bond (having no amide hydrogen), and also because its sidechain interferes sterically with the backbone of the preceding turn - inside a helix, this forces a bend of about 30° in the helix axis. However, proline is often seen as the first residue of a helix, it is presumed due to its structural rigidity. At the other extreme, glycine also tends to disrupt helices because its high conformational flexibility makes it entropically expensive to adopt the relatively constrained α-helical structure.

Table of Standard Amino Acid Alpha-Helical Propensities

Estimated differences in free energy, $\Delta(\Delta G)$, estimated in kcal/mol per residue in an alpha-helical configuration, relative to Alanine arbitrarily set as zero. Higher numbers (more positive free energies) are less favoured. Significant deviations from these average numbers are possible, depending on the identities of the neighbouring residues.

Amino Acid	3-Letter	1-Letter	Helical Penalty (Kcal/mol)
Alanine	Ala	A	0
Arginine	Arg	R	0.21
Asparagine	Asn	N	0.65
Aspartic acid	Asp	D	0.69
Cysteine	Cys	C	0.68
Glutamic acid	Glu	E	0.40
Glutamine	Gln	Q	0.39
Glycine	Gly	G	1.00
Histidine	His	H	0.61
Isoleucine	Ile	I	0.41
Leucine	Leu	L	0.21
Lysine	Lys	K	0.26
Methionine	Met	M	0.24
Phenylalanine	Phe	F	0.54
Proline	Pro	P	3.16
Serine	Ser	S	0.50
Threonine	Thr	T	0.66
Tryptophan	Trp	W	0.49
Tyrosine	Tyr	Y	0.53
Valine	Val	V	0.61

Dipole Moment

A helix has an overall dipole moment due to the aggregate effect of the individual microdipoles from the carbonyl groups of the peptide bond pointing along the helix axis. The effects of this macrodipole are a matter of some controversy. α-helices often occur with the N-terminal end bound by a negatively charged group, sometimes an amino acid side chain such as glutamate or aspartate, or sometimes a phosphate ion. Some regard the helix macrodipole as interacting electrostatically with such groups. Others feel that this is misleading and it is more realistic to say that the hydrogen bond potential of the free NH groups at the N-terminus of an α-helix can be satisfied by hydrogen bonding; this can also be regarded as set of interactions between local microdipoles such as C=O...H-N.

Coiled-Coils

Coiled-coil α helices are highly stable forms in which two or more helices wrap around each other in a "supercoil" structure. Coiled coils contain a highly characteristic sequence motif known as a heptad repeat, in which the motif repeats itself every seven residues along the sequence (amino acid residues, not DNA base-pairs). The first and especially the fourth residues (known as the a and d positions) are almost always hydrophobic; the fourth residue is typically leucine - this gives rise to the name of the structural motif called a leucine zipper, which is a type of coiled-coil. These hydrophobic residues and pack together in the interior of the helix bundle. In general, the fifth and seventh residues (the e and g positions) have opposing charges and form a salt bridge stabilized by electrostatic interactions. Fibrous proteins such as keratin or the "stalks" of myosin or kinesin often adopt coiled-coil structures, as do several dimerizing proteins. A pair of coiled-coils - a four-helix bundle - is a very common structural motif in proteins. For example, it occurs in human growth hormone and several varieties of cytochrome. The Rop protein, which promotes plasmid replication in bacteria, is an interesting case in which a single polypeptide forms a coiled-coil and two monomers assemble to form a four-helix bundle.

The amino acids that make up a particular helix can be plotted on a helical wheel, a representation that illustrates the orientations of the constituent amino acids. Often in globular proteins, as well as in specialized structures such as coiled-coils and leucine zippers, an α-helix will exhibit two "faces" - one containing predominantly hydrophobic amino acids oriented toward the interior of the protein, in the hydrophobic core, and one containing predominantly polar amino acids oriented toward the solvent-exposed surface of the protein.

Larger-Scale Assemblies

Myoglobin and hemoglobin, the first two proteins whose structures were solved by X-ray crystallography, have very similar folds made up of about 70% α-helix, with the rest being non-repetitive regions, or "loops" that connect the helices. In classifying proteins by their dominant fold, the Structural Classification of Proteins database maintains a large category specifically for all-α proteins.

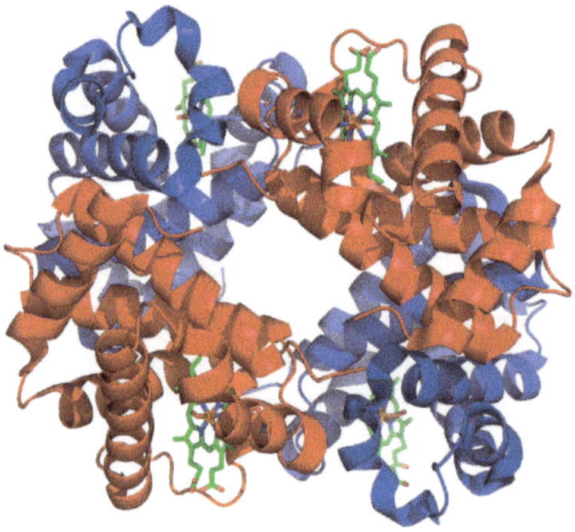

The Hemoglobin molecule has four heme-binding subunits, each made largely of α-helices.

Hemoglobin then has an even larger-scale quaternary structure, in which the functional oxygen-binding molecule is made up of four subunits.

Functional Roles

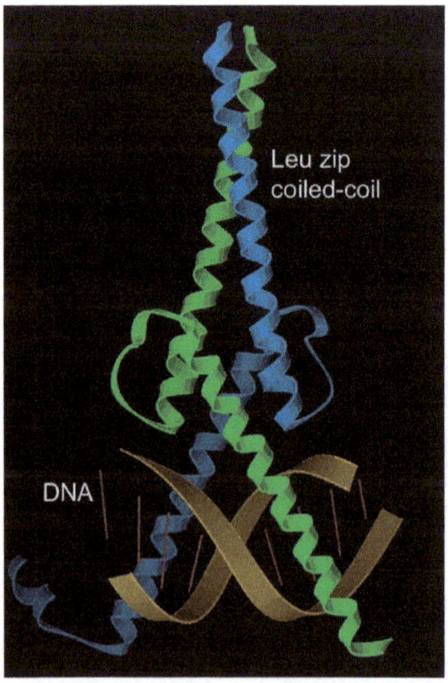

Leucine zipper coiled-coil helices & DNA-binding helices: transcription factor Max (PDB file 1HLO)

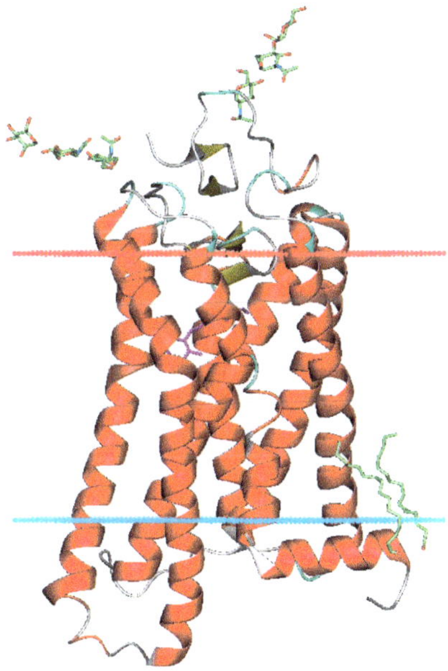

Bovine rhodopsin (PDB file 1GZM), with a bundle of seven helices crossing the membrane (membrane surfaces marked by horizontal lines)

DNA Binding

α-Helices have particular significance in DNA binding motifs, including helix-turn-helix motifs, leucine zipper motifs and zinc finger motifs. This is because of the convenient structural fact that the diameter of an α-helix is about 12Å (1.2 nm) including an average set of sidechains, about the same as the width of the major groove in B-form DNA, and also because coiled-coil (or leucine zipper) dimers of helices can readily position a pair of interaction surfaces to contact the sort of symmetrical repeat common in double-helical DNA. An example of both aspects is the transcription factor Max, which uses a helical coiled-coil to dimerize, positioning another pair of helices for interaction in two successive turns of the DNA major groove.

Membrane Spanning

α-Helices are also the most common protein structure element that crosses biological membranes (transmembrane protein) it is presumed because the helical structure can satisfy all backbone hydrogen-bonds internally, leaving no polar groups exposed to the membrane if the sidechains are hydrophobic. Proteins are sometimes anchored by a single membrane-spanning helix, sometimes by a pair, and sometimes by a helix bundle, most classically consisting of seven helices arranged up-and-down in a ring such as for rhodopsins or for [G protein–coupled receptor]s (GPCRs).

Mechanical Properties

α-Helices under axial tensile deformation, a characteristic loading condition that appears in many alpha-helix-rich filaments and tissues, results in a characteristic three-phase behavior of stiff-soft-stiff tangent modulus. Phase I corresponds to the small-deformation regime during which the helix is stretched homogeneously, followed by phase II, in which alpha-helical turns break mediated by the rupture of groups of H-bonds. Phase III is typically associated with large-deformation covalent bond stretching.

Dynamical Features

Alpha-helices in proteins may have low-frequency accordion-like motion as observed by the Raman spectroscopy and analyzed via the quasi-continuum model. Helices not stabilized by tertiary interactions show dynamic behavior, which can be mainly attributed to helix fraying from the ends.

Helix-Coil Transition

Homopolymers of amino-acids (such as poly-lysine) can adopt α-helical structure at low temperature that is "melted out" at high temperatures. This helix-coil transition was once thought to be analogous to protein denaturation. The statistical mechanics of this transition can be modeled using an elegant transfer matrix method, characterized by two parameters: the propensity to initiate a helix and the propensity to extend a helix.

The α-Helix In Art

Julian Voss-Andreae's *Alpha Helix for Linus Pauling* (2004), powder coated steel, height 10 ft (3 m). The sculpture stands in front of Pauling's childhood home on 3945 SE Hawthorne Boulevard in Portland, Oregon, USA.

At least five artists have made explicit reference to the α-helix in their work: Julie Newdoll in painting and Julian Voss-Andreae, Bathsheba Grossman, Byron Rubin, and Mike Tyka in sculpture.

San Francisco area artist Julie Newdoll, who holds a degree in Microbiology with a minor in art, has specialized in paintings inspired by microscopic images and molecules since 1990. Her painting "Rise of the Alpha Helix" (2003) features human figures arranged in an α helical arrangement. According to the artist, "the flowers reflect the various types of sidechains that each amino acid holds out to the world". It is interesting to note that this same metaphor is also echoed from the scientist's side: "β sheets do not show a stiff repetitious regularity but flow in graceful, twisting curves, and even the α-helix is regular more in the manner of a flower stem, whose branching nodes show the influence of environment, developmental history, and the evolution of each part to match its own idiosyncratic function."

Julian Voss-Andreae is a German-born sculptor with degrees in experimental physics and sculpture. Since 2001 Voss-Andreae creates "protein sculptures" based on protein structure with the α-helix being one of his preferred objects. Voss-Andreae has made α-helix sculptures from diverse materials including bamboo and whole trees. A monument Voss-Andreae created in 2004 to celebrate the memory of Linus Pauling, the discoverer of the α-helix, is fashioned from a large steel beam rearranged in the structure of the α-helix. The 10-foot (3 m) tall, bright-red sculpture stands in front of Pauling's childhood home in Portland, Oregon.

Ribbon diagrams of α-helices are a prominent element in the laser-etched crystal sculptures of protein structures created by artist Bathsheba Grossman, such as those of insulin, hemoglobin, and DNA polymerase. Byron Rubin is a former protein crystallographer now professional sculptor in metal of proteins, nucleic acids, and drug molecules - many of which featuring α-helices, such as subtilisin, human growth hormone, and phospholipase A2.

Mike Tyka is a computational biochemist at the University of Washington working with David Baker. Tyka has been making sculptures of protein molecules since 2010 from copper and steel, including ubiquitin and a potassium channel tetramer.

Beta Sheet

The β-sheet (also β-pleated sheet) is a common motif of regular secondary structure in proteins. Beta sheets consist of beta strands (also β-strand) connected laterally by at least two or three backbone hydrogen bonds, forming a generally twisted, pleated sheet. A β-strand is a stretch of polypeptide chain typically 3 to 10 amino acids long with backbone in an extended conformation. The supramolecular association of β-sheets has been implicated in formation of the protein aggregates and fibrils observed in many human diseases, notably the amyloidoses such as Alzheimer's disease.

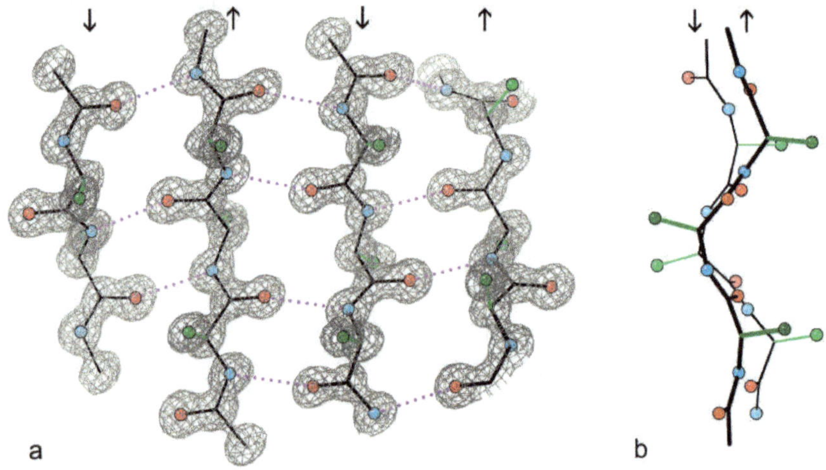

An example of a 4-stranded antiparallel β-sheet fragment from a crystal structure of the enzyme catalase (PDB file 1GWE at 0.88Å resolution). a) Front view, showing the antiparallel hydrogen bonds (dotted) between peptide NH and CO groups on adjacent strands. Arrows indicate chain direction, and electron density contours outline the non-H atoms. O atoms are red balls, N atoms are blue, and H atoms are omitted for simplicity; sidechains are shown only out to the first sidechain C atom (green). b) Edge-on view of the central two β-strands in a, showing the righthanded *twist* and the *pleat* of Cαs and sidechains that alternately stick out in opposite directions from the sheet.

The first β-sheet structure was proposed by William Astbury in the 1930s. He proposed the idea of hydrogen bonding between the peptide bonds of parallel or antiparallel extended β-strands. However, Astbury did not have the necessary data on the bond geometry of the amino acids in order to build accurate models, especially since he did not then know that the peptide bond was planar. A refined version was proposed by Linus Pauling and Robert Corey in 1951. Their model incorporated the planarity of the peptide bond which they previously explained as resulting from keto-enol tautomerization.

Structure and Orientation

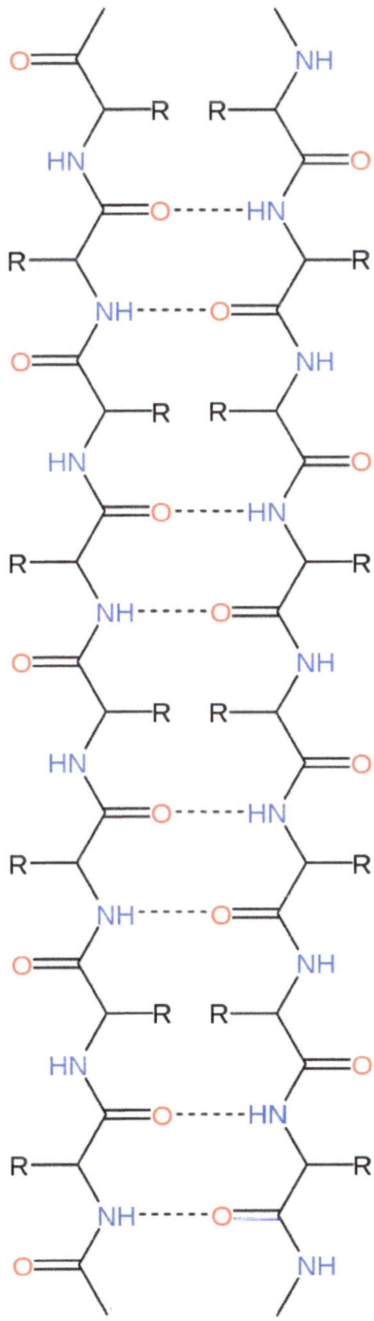

Antiparallel β-sheet hydrogen bonding patterns, represented by dotted lines. Oxygen atoms are colored red and nitrogen atoms colored **blue**.

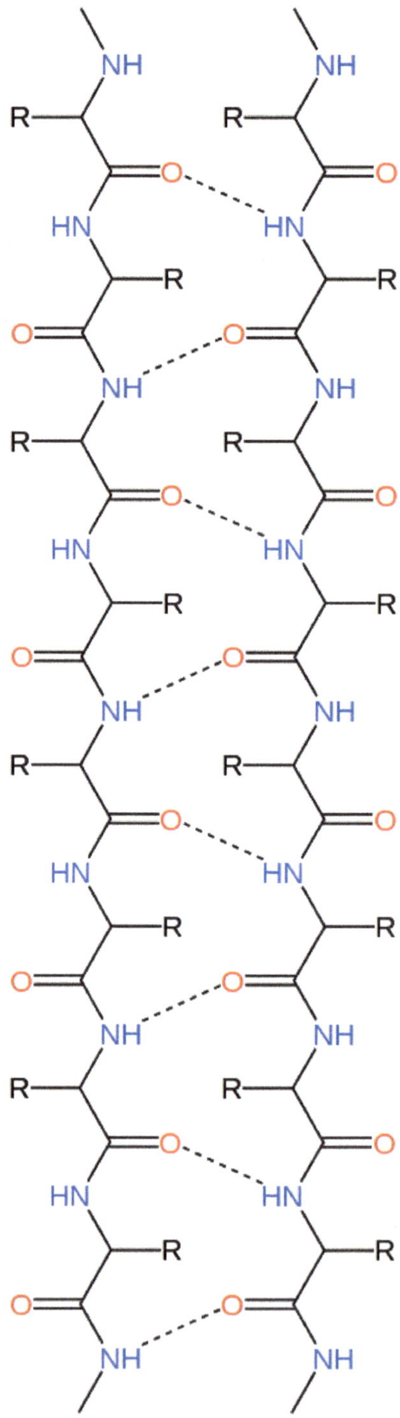

Parallel β-sheet hydrogen bonding patterns, represented by dotted lines. Oxygen atoms are colored red and nitrogen atoms colored **blue**.

Geometry

The majority of β-strands are arranged adjacent to other strands and form an extensive hydrogen bond network with their neighbors in which the N-H groups in the backbone of one strand establish hydrogen bonds with the C=O groups in the backbone of the adjacent strands. In the fully extended β-strand, successive side chains point straight up, then straight down, then straight up, etc. Adjacent β-strands in a β-sheet are aligned so that their Cα atoms are adjacent and their side chains point in the same direction. The "pleated" appearance of β-strands arises from tetrahedral chemical bonding at the Cα atom; for example, if a side chain points straight up, then the bond to the C' must point slightly downwards, since its bond angle is approximately 109.5°. The pleating causes the distance between $C^{\alpha}{}_{i}$ and C^{α}_{i+2} to be approximately 6 Å, rather than the 7.6 Å (2 × 3.8 Å) expected from two fully extended trans peptides. The "sideways" distance between adjacent Cα atoms in hydrogen-bonded β-strands is roughly 5 Å.

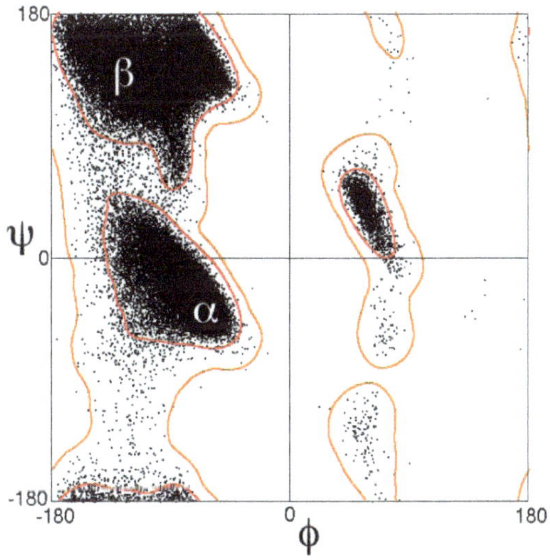

Ramachandran (φ,ψ) plot of about 100,000 high-resolution datapoints, showing the broad, favorable region around the conformation typical for β-sheet residues.

However, β-strands are rarely perfectly extended; rather, they exhibit a twist due to the chirality of their component amino acids. The energetically preferred dihedral angles near (φ, ψ) = (–135°, 135°) (broadly, the upper left region of the Ramachandran plot) diverge significantly from the fully extended conformation (φ, ψ) = (–180°, 180°). The twist is often associated with alternating fluctuations in the dihedral angles to prevent the individual β-strands in a larger sheet from splaying apart. A good example of a strongly twisted β-hairpin can be seen in the protein BPTI.

The side chains point outwards from the folds of the pleats, roughly perpendicularly to the plane of the sheet; successive residues point outwards on alternating faces of the sheet.

Hydrogen Bonding Patterns

Because peptide chains have a directionality conferred by their N-terminus and C-terminus, β-strands too can be said to be directional. They are usually represented in protein topology diagrams by an arrow pointing toward the C-terminus. Adjacent β-strands can form hydrogen bonds in antiparallel, parallel, or mixed arrangements.

In an antiparallel arrangement, the successive β-strands alternate directions so that the N-terminus of one strand is adjacent to the C-terminus of the next. This is the arrangement that produces the strongest inter-strand stability because it allows the inter-strand hydrogen bonds between carbonyls and amines to be planar, which is their preferred orientation. The peptide backbone dihedral angles (φ, ψ) are about (–140°, 135°) in antiparallel sheets. In this case, if two atoms C_i^α and C_j^α are adjacent in two hydrogen-bonded β-strands, then they form two mutual backbone hydrogen bonds to each other's flanking peptide groups; this is known as a close pair of hydrogen bonds.

In a parallel arrangement, all of the N-termini of successive strands are oriented in the same direction; this orientation may be slightly less stable because it introduces nonplanarity in the inter-strand hydrogen bonding pattern. The dihedral angles (φ, ψ) are about (–120°, 115°) in parallel sheets. It is rare to find less than five interacting parallel strands in a motif, suggesting that a smaller number of strands may be unstable, however it is also fundamentally more difficult for parallel β-sheets to form because strands with N and C termini aligned necessarily must be very distant in sequence. There is also evidence that parallel β-sheet may be more stable since small amyloidogenic sequences appear to generally aggregate into β-sheet fibrils composed of primarily parallel β-sheet strands, where one would expect anti-parallel fibrils if anti-parallel were more stable.

In parallel β-sheet structure, if two atoms C_i^α and C_j^α are adjacent in two hydrogen-bonded β-strands, then they do not hydrogen bond to each other; rather, one residue forms hydrogen bonds to the residues that flank the other (but not vice versa). For example, residue i may form hydrogen bonds to residues $j-1$ and $j+1$; this is known as a wide pair of hydrogen bonds. By contrast, residue j may hydrogen-bond to different residues altogether, or to none at all.

Finally, an individual strand may exhibit a mixed bonding pattern, with a parallel strand on one side and an antiparallel strand on the other. Such arrangements are less common than a random distribution of orientations would suggest, suggesting that this pattern is less stable than the anti-parallel arrangement, however bioinformatic analysis always struggles with extracting structural thermodynamics since there are always numerous other structural features present in whole proteins. Also proteins are inherently constrained by folding kinetics as well as folding thermodynamics, so one must always be careful in concluding stability from bioinformatic analysis.

The hydrogen bonding of β-strands need not be perfect, but can exhibit localized disruptions known as β-bulges.

The hydrogen bonds lie roughly in the plane of the sheet, with the peptide carbonyl groups pointing in alternating directions with successive residues; for comparison, successive carbonyls point in the same direction in the alpha helix.

Amino Acid Propensities

Large aromatic residues (Tyr, Phe and Trp) and β-branched amino acids (Thr, Val, Ile) are favored to be found in β-strands in the middle of β-sheets. Interestingly, different types of residues (such as Pro) are likely to be found in the edge strands in β-sheets, presumably to avoid the "edge-to-edge" association between proteins that might lead to aggregation and amyloid formation.

Common Structural Motifs

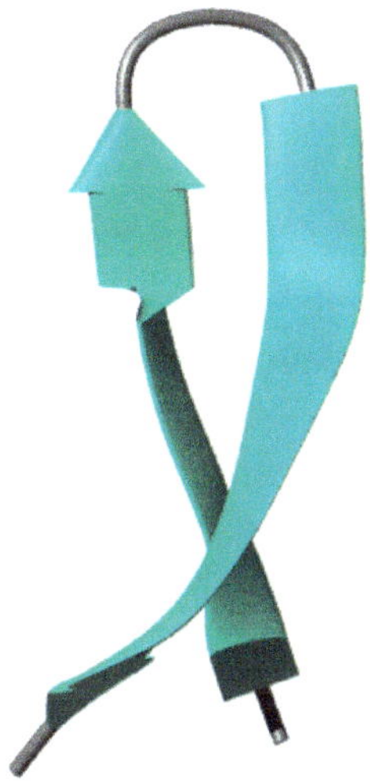

The β-hairpin motif

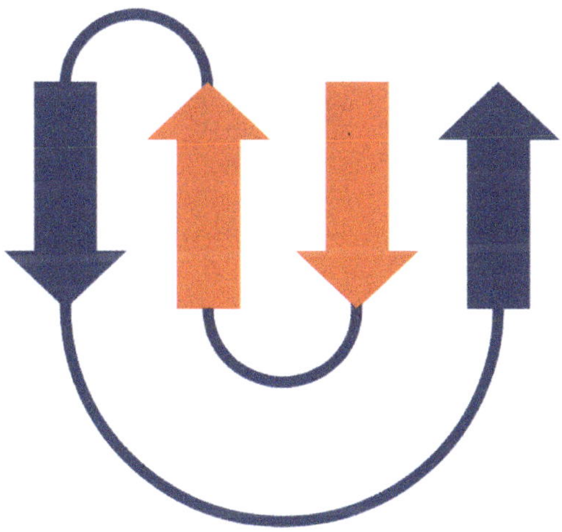

The Greek-key motif

A very simple structural motif involving β-sheets is the β-hairpin, in which two antiparallel strands are linked by a short loop of two to five residues, of which one is frequently a glycine or a proline, both of which can assume the dihedral-angle conformations required for a tight turn or a β-bulge loop. Individual strands can also be linked in more elaborate ways with longer loops that may contain α-helices.

Greek Key Motif

The Greek key motif consists of four adjacent antiparallel strands and their linking loops. It consists of three antiparallel strands connected by hairpins, while the fourth is adjacent to the first and linked to the third by a longer loop. This type of structure forms easily during the protein folding process. It was named after a pattern common to Greek ornamental artwork.

The β-α-β Motif

Due to the chirality of their component amino acids, all strands exhibit right handed twist evident in most higher-order β-sheet structures. In particular, the linking loop between two parallel strands almost always has a right-handed crossover chirality, which is strongly favored by the inherent twist of the sheet. This linking loop frequently contains a helical region, in which case it is called a β-α-β motif. A closely related motif called a β-α-β-α motif forms the basic component of the most commonly observed protein tertiary structure, the TIM barrel.

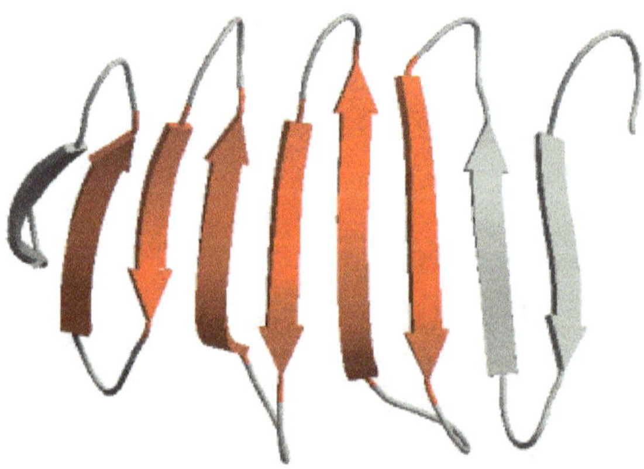

The β-meander motif from the outer surface Protein A

β-Meander Motif

A simple supersecondary protein topology composed of 2 or more consecutive antiparallel β-strands linked together by hairpin loops. This motif is common in β-sheets and can be found in several structural architectures including β-barrels and β-propellers.

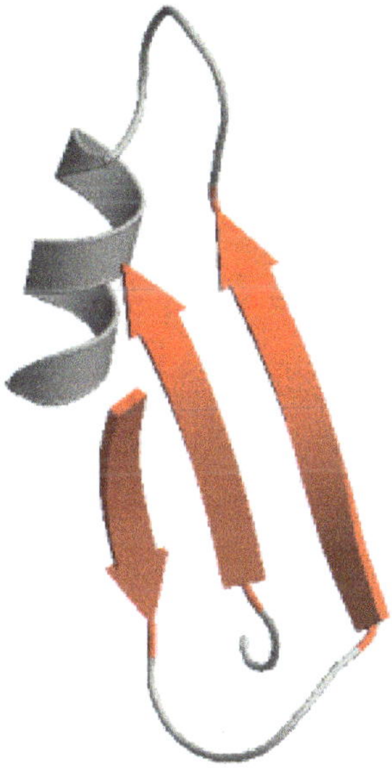

Psi-loop motif from Carboxypeptidase A

Psi-Loop Motif

The psi-loop (Ψ-loop) motif consists of two antiparallel strands with one strand in between that is connected to both by hydrogen bonds.There are four possible strand topologies for single Ψ-loops. This motif is rare as the process resulting in its formation seems unlikely to occur during protein folding. The Ψ-loop was first identified in the aspartic protease family.

Structural Architectures of Proteins with β-Sheets

β-sheets are present in all-β, α+β and α/β domains, and in many peptides or small proteins with poorly defined overall architecture. All-β domains may form β-barrels, β-sandwiches, β-prisms, β-propellers, and β-helices.

Structural Topology

The topology of a β-sheet describes the order of hydrogen-bonded β-strands along the backbone. For example, the flavodoxin fold has a five-stranded, parallel β-sheet with topology 21345; thus, the edge strands are β-strand 2 and β-strand 5 along the backbone. Spelled out explicitly, β-strand 2 is H-bonded to β-strand 1, which is H-bonded to β-strand 3, which is H-bonded to β-strand 4, which is H-bonded to β-strand 5, the other edge strand. In the same system, the Greek key motif described above has a 4123 topology.

The secondary structure of a β-sheet can be described roughly by giving the number of strands, their topology, and whether their hydrogen bonds are parallel or antiparallel.

β-sheets can be open, meaning that they have two edge strands (as in the flavodoxin fold or the immunoglobulin fold) or they can be closed β-barrels (such as the TIM barrel). β-Barrels are often described by their stagger or shear. Some open β-sheets are very curved and fold over on themselves (as in the SH3 domain) or form horseshoe shapes (as in the ribonuclease inhibitor). Open β-sheets can assemble face-to-face (such as the β-propeller domain or immunoglobulin fold) or edge-to-edge, forming one big β-sheet.

Dynamic Features

β-pleated sheet structures are made from extended β-strand polypeptide chains, with strands linked to their neighbours by hydrogen bonds. Due to this extended backbone conformation, β-sheets resist Stretching. β-sheets in proteins may carry out low-frequency accordion-like motion as observed by the Raman spectroscopy and analyzed with the quasi-continuum model.

Parallel β-Helices

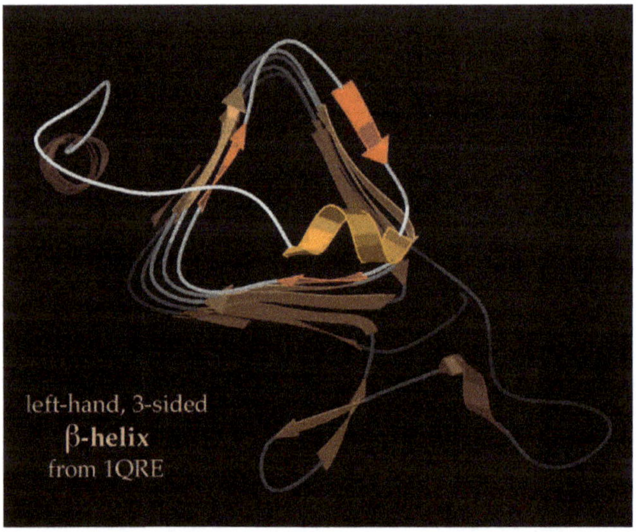

End-view of a 3-sided, left handed β-helix (PDB: 1QRE)

A β-helix is formed from repeating structural units consisting of two or three short β-strands linked by short loops. These units "stack" atop one another in a helical fashion so that successive repetitions of the same strand hydrogen-bond with each other in a parallel orientation.

In left-handed β-helices, the strands themselves are quite straight and untwisted; the resulting helical surfaces are nearly flat, forming a regular triangular prism shape, as shown for the 1QRE archaeal carbonic anhydrase at right. Other examples are the lipid A synthesis enzyme LpxA and insect antifreeze proteins with a regular array of Thr sidechains on one face that mimic the structure of ice.

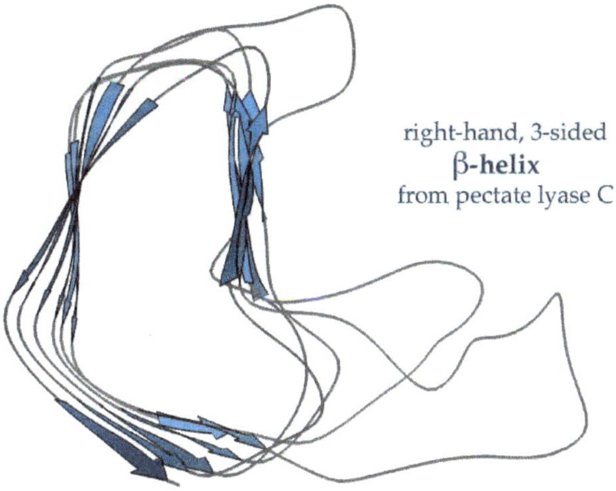

End-view of a 3-sided, right handed β-helix.

Righthanded β-helices, typified by the pectate lyase enzyme shown at left or P22 phage tailspike protein, have a less regular cross-section, longer and indented on one of the sides; of the three linker loops, one is consistently just two residues long and the others are variable, often elaborated to form a binding or active site.

A two-sided β-helix (right-handed) is found in some bacterial metalloproteases; its two loops are each six residues long and bind stabilizing calcium ions to maintain the integrity of the structure, using the backbone and the Asp side chain oxygens of a GGXGXD sequence motif. This fold is called a β-roll in the SCOP classification.

β-Sheets in Pathology

Some proteins that are disordered or helical as monomers, such as amyloid β (see amyloid plaque) can form β-sheet-rich oligomeric structures associated with pathological states. The amyloid β protein's oligomeric form is implicated as a cause of Alzheimer's. Its structure has yet to be determined in full, but recent data suggest that it may resemble an unusual two-strand β-helix.

The side chains from the amino acid residues found in a β-sheet structure may also be arranged such that many of the adjacent sidechains on one side of the sheet are hydrophobic, while many of those adjacent to each other on the alternate side of the sheet are polar or charged (hydrophilic), which can be useful if the sheet is to form a boundary between polar/watery and nonpolar/greasy environments.

Loop

Loops are regions of a protein chain that are (1) between α helices and β sheets, (2) of various lengths and three-dimensional configurations, and (3) on the surface of the structure. Hairpin loops that represent a complete turn in the polypeptide chain joining two antiparallel β strands may be as short as two amino acids in length. Loops interact with the surrounding aqueous environment and other proteins. Because amino acids in loops are not constrained by space and environment as are amino acids in the core region, and do not have an effect on the arrangement of secondary structures in the core, more substitutions, insertions, and deletions may occur. Thus, in a sequence alignment, the presence of these features may be an indication of a loop. The positions of introns in genomic DNA sometimes correspond to the locations of loops in the encoded protein. Loops also tend to have charged and polar amino acids and are frequently a component of active sites. A detailed examination of loop structures has shown that they fall into distinct families.

Coils

A region of secondary structure that is not a α helix, a β sheet, or a recognizable turn is commonly referred to as a coil.

REFERENCES

1. Mount DM (2004). Bioinformatics: Sequence and Genome Analysis. 2. Cold Spring Harbor Laboratory Press. ISBN 0-87969-712-1.
2. Huang, JY.; Brutlag, DL. (Jan 2001). "The EMOTIF database". Nucleic Acids Res. 29 (1): 202–04. doi:10.1093/nar/29.1.202. PMID 11125091.
3. Pirovano, W.; Heringa, J. (2010). "Protein secondary structure prediction". Methods Mol Biol. Methods in Molecular Biology. 609: 327–48. doi:10.1007/978-1-60327-241-4_19. ISBN 978-1-60327-240-7. PMID 20221928.
4. Guzzo, AV (1965). "Influence of Amino-Acid Sequence on Protein Structure". Biophys. J. 5 (6): 809–822. Bibcode:1965BpJ.....5..809G. doi:10.1016/S0006-3495(65)86753-4. PMC 1367904Freely accessible. PMID 5884309.
5. Prothero, JW (1966). "Correlation between Distribution of Amino Acids and Alpha Helices". Biophys. J. 6 (3): 367–370. Bibcode:1966BpJ.....6..367P. doi:10.1016/S0006-3495(66)86662-6. PMC 1367951Freely accessible. PMID 5962284.
6. Schiffer, M; Edmundson AB (1967). "Use of Helical Wheels to Represent Structures of Proteins and to Identify Segments with Helical Potential". Biophys. J. 7 (2): 121–35. Bibcode:1967BpJ.....7..121S. doi:10.1016/S0006-3495(67)86579-2. PMC 1368002Freely accessible. PMID 6048867.
7. Kotelchuck, D; Scheraga HA (1969). "The Influence of Short-Range Interactions on Protein Conformation, II. A Model for Predicting the α-Helical Regions of Proteins". Proc Natl Acad Sci USA. 62 (1): 14–21. doi:10.1073/pnas.62.1.14. PMC 285948Freely accessible. PMID 5253650.
8. Lewis, PN; Gō N; Gō M; Kotelchuck D; Scheraga HA (1970). "Helix Probability Profiles of Denatured Proteins and Their Correlation with Native Structures". Proc Natl Acad Sci USA. 65 (4): 810–5. doi:10.1073/pnas.65.4.810. PMC 282987Freely accessible. PMID 5266152.

9. Froimowitz M, Fasman GD; Fasman (1974). "Prediction of the secondary structure of proteins using the helix-coil transition theory". Macromolecules. 7 (5): 583–9. doi:10.1021/ma60041a009. PMID 4371089.

10. Dor O, Zhou Y; Zhou (2006). "Achieving 80% tenfold cross-validated accuracy for secondary structure prediction by large-scale training". Proteins. 66 (4): 838–45. doi:10.1002/prot.21298. PMID 17177203.

11. Chou PY, Fasman GD; Fasman (1974). "Prediction of protein conformation". Biochemistry. 13 (2): 222–245. doi:10.1021/bi00699a002. PMID 4358940.

12. Garnier J, Osguthorpe DJ, Robson B; Osguthorpe; Robson (1978). "Analysis of the accuracy and implications of simple methods for predicting the secondary structure of globular proteins". J Mol Biol. 120 (1): 97–120. doi:10.1016/0022-2836(78)90297-8. PMID 642007.

13. Pham TH, Satou K, Ho TB; Satou; Ho (2005). "Support vector machines for prediction and analysis of beta and gamma-turns in proteins". J Bioinform Comput Biol. 3 (2): 343–358. doi:10.1142/S0219720005001089. PMID 15852509.

14. Zhang Q, Yoon S, Welsh WJ; Yoon; Welsh (2005). "Improved method for predicting beta-turn using support vector machine". Bioinformatics. 21 (10): 2370–4. doi:10.1093/bioinformatics/bti358. PMID 15797917.

15. Zimmermann O, Hansmann UH; Hansmann (2006). "Support vector machines for prediction of dihedral angle regions". Bioinformatics. 22 (24): 3009–15. doi:10.1093/bioinformatics/btl489. PMID 17005536.

16. Kuang R, Leslie CS, Yang AS; Leslie; Yang (2004). "Protein backbone angle prediction with machine learning approaches". Bioinformatics. 20 (10): 1612–21. doi:10.1093/bioinformatics/bth136. PMID 14988121.

17. Faraggi E, Yang Y, Zhou Y; Yang; Zhang; Zhou (2009). "Predicting continuous local structure and the effect of its substitution for secondary structure in fragment-free protein structure prediction". Structure. 17 (11): 1515–1527. doi:10.1016/j.str.2009.09.006. PMC 2778607Freely accessible. PMID 19913486.

18. Zhong L, Johnson WC Jr; Johnson Jr (1992). "Environment affects amino acid preference for secondary structure". Proc Natl Acad Sci USA. 89 (10): 4462–5. doi:10.1073/pnas.89.10.4462. PMC 49102Freely accessible. PMID 1584778.

19. Macdonald JR, Johnson WC Jr; Johnson Jr (2001). "Environmental features are important in determining protein secondary structure". Protein Sci. 10 (6): 1172–7. doi:10.1110/ps.420101. PMC 2374018Freely accessible. PMID 11369855.

20. Costantini S, Colonna G, Facchiano AM; Colonna; Facchiano (2006). "Amino acid propensities for secondary structures are influenced by the protein structural class". Biochem Biophys Res Commun. 342 (2): 441–451. doi:10.1016/j.bbrc.2006.01.159. PMID 16487481.

21. Marashi SA; et al. (2007). "Adaptation of proteins to different environments: a comparison of proteome structural properties in Bacillus subtilis and Escherichia coli". J Theor Biol. 244 (1): 127–132. doi:10.1016/j.jtbi.2006.07.021. PMID 16945389.

22. Costantini S, Colonna G, Facchiano AM; Colonna; Facchiano (2007). "PreSSAPro: a software for the prediction of secondary structure by amino acid properties". Comput Biol Chem. 31 (5–6): 389–392. doi:10.1016/j.compbiolchem.2007.08.010. PMID 17888742.

23. Momen-Roknabadi A; et al. (2008). "Impact of residue accessible surface area on the prediction of protein secondary structures". BMC Bioinformatics. 9: 357. doi:10.1186/1471-2105-9-357. PMC 2553345Freely accessible. PMID 18759992.

24. Adamczak R, Porollo A, Meller J; Porollo; Meller (2005). "Combining prediction of secondary structure and solvent accessibility in proteins". Proteins. 59 (3): 467–475. doi:10.1002/prot.20441. PMID 15768403.

25. Lakizadeh A, Marashi SA (2009). "Addition of contact number information can improve protein secondary structure prediction by neural networks" (PDF). Excli J. 8: 66–73.

26. Zhang Y (2008). "Progress and challenges in protein structure prediction". Curr Opin Struct Biol. 18 (3): 342–8. doi:10.1016/j.sbi.2008.02.004. PMC 2680823Freely accessible. PMID 18436442.

27. Shaw, David E.; Bowers, Kevin J.; Chow, Edmond; Eastwood, Michael P.; Ierardi, Douglas J.; Klepeis, John L.; Kuskin, Jeffrey S.; Larson, Richard H.; Lindorff-Larsen, Kresten; Maragakis, Paul; Moraes, Mark A.; Dror, Ron O.; Piana, Stefano; Shan, Yibing; Towles, Brian; Salmon, John K.; Grossman, J. P.; Mackenzie, Kenneth M.; Bank, Joseph A.; Young, Cliff; Deneroff, Martin M.; Batson, Brannon (2009). "Proceedings of the Conference on High Performance Computing Networking, Storage and Analysis - SC '09": 1. doi:10.1145/1654059.1654126. ISBN 9781605587448. |chapter= ignored (help)

28. Pierce, Levi C.T.; Salomon-Ferrer, Romelia; Augusto F. de Oliveira, Cesar; McCammon, J. Andrew; Walker, Ross C. (2012). "Routine Access to Millisecond Time Scale Events with Accelerated Molecular Dynamics". Journal of Chemical Theory and Computation. 8 (9): 2997–3002. doi:10.1021/ct300284c. ISSN 1549-9618. PMID 22984356.

29. Kmiecik, Sebastian; Gront, Dominik; Kolinski, Michal; Wieteska, Lukasz; Dawid, Aleksandra Elzbieta; Kolinski, Andrzej (2016-06-22). "Coarse-Grained Protein Models and Their Applications". Chemical Reviews. doi:10.1021/acs.chemrev.6b00163. ISSN 0009-2665.

30. Gobel, U.; et al. (1994). "Correlated mutations and residue contacts in proteins". Proteins. 18 (4): 309–17. doi:10.1002/prot.340180402. PMID 8208723.

31. Taylor, W. R.; Hatrick, K. (1994). "Compensating changes in protein multiple sequence alignments". Protein Eng. 7 (3): 341–48. doi:10.1093/protein/7.3.341. PMID 8177883.

32. Neher, E. (Jan 1994). "How frequent are correlated changes in families of protein sequences?". Proc Natl Acad Sci U S A. 91 (1): 98–102. doi:10.1073/pnas.91.1.98. PMC 42893Freely accessible. PMID 8278414.

33. Marks, D. S.; et al. (2011). "Protein 3D structure computed from evolutionary sequence variation". PLoS ONE. 6 (12): e28766. doi:10.1371/journal.pone.0028766. PMC 3233603Freely accessible. PMID 22163331.
34. Burger, L.; van Nimwegen, E (2010). "Disentangling direct from indirect co-evolution of residues in protein alignments". PLoS Comput Biol. 6: e1000633. doi:10.1371/journal.pcbi.1000633.
35. Morcos, F.; Pagnani, A.; Lunt, B.; Bertolino, A.; Marks, DS.; Sander, C.; Zecchina, R.; Onuchic, JN.; et al. (Dec 2011). "Direct-coupling analysis of residue coevolution captures native contacts across many protein families". Proc Natl Acad Sci U S A. 108 (49): E1293–1301. doi:10.1073/pnas.1111471108. PMC 3241805Freely accessible. PMID 22106262.
36. Nugent, T.; Jones, D.T. (2012). "Accurate de novo structure prediction of large transmembrane protein domains using fragment-assembly and correlated mutation analysis". Proc Natl Acad Sci U S A. 109 (24): E1540–47. doi:10.1073/pnas.1120036109. PMID 22645369.
37. Hopf, T. A.; et al. (2012). "Three-dimensional structures of membrane proteins from genomic sequencing". Cell. 149 (7): 1607–21. doi:10.1016/j.cell.2012.04.012. PMC 3641781Freely accessible. PMID 22579045.
38. Zhang Y & Skolnick J (2005). "The protein structure prediction problem could be solved using the current PDB library". Proc Natl Acad Sci USA. 102 (4): 1029–34. doi:10.1073/pnas.0407152101. PMC 545829Freely accessible. PMID 15653774.
39. Bowie JU, Luthy R, Eisenberg D; Lüthy; Eisenberg (1991). "A method to identify protein sequences that fold into a known three-dimensional structure". Science. 253 (5016): 164–170. doi:10.1126/science.1853201. PMID 1853201.
40. Dunbrack, RL (2002). "Rotamer Libraries in the 21st Century". Curr. Opin. Struct. Biol. 12 (4): 431–440. doi:10.1016/S0959-440X(02)00344-5. PMID 12163064.
41. Ponder JW, Richards FM; Richards (1987). "Tertiary templates for proteins: use of packing criteria in the enumeration of allowed sequences for different structural classes". J. Mol. Biol. 193 (4): 775–791. doi:10.1016/0022-2836(87)90358-5. PMID 2441069.
42. Lovell SC, Word, Richardson, Richardson (2000). "The penultimate rotamer library". Proteins: Struc. Func. Genet. 40 (3): 389–408. doi:10.1002/1097-0134(20000815)40:3<389::AID-PROT50>3.0.CO;2-2.
43. Shapovalov MV, Dunbrack, RL; Dunbrack Jr (2011). "A smoothed backbone-dependent rotamer library for proteins derived from adaptive kernel density estimates and regressions". Structure (Cell Press). 19 (6): 844–858. doi:10.1016/j.str.2011.03.019. PMC 3118414Freely accessible. PMID 21645855.
44. Chen, VB.; Arendall, WB.; Headd, JJ.; Keedy, DA.; Immormino, RM.; Kapral, GJ.; Murray, LW.; Richardson, JS.; Richardson, DC. (Jan 2010). "MolProbity: all-atom structure validation for macromolecular crystallography". Acta Crystallogr D. 66 (Pt 1): 12–21. doi:10.1107/S0907444909042073. PMC 2803126Freely accessible. PMID 20057044.

45. Bower, MJ.; Cohen, FE.; Dunbrack, RL. (Apr 1997). "Prediction of protein side-chain rotamers from a backbone-dependent rotamer library: a new homology modeling tool". J Mol Biol. 267 (5): 1268–82. doi:10.1006/jmbi.1997.0926. PMID 9150411.

46. Voigt CA, Gordon DB, Mayo SL; Gordon; Mayo (2000). "Trading accuracy for speed: A quantitative comparison of search algorithms in protein sequence design". J Mol Biol. 299 (3): 789–803. doi:10.1006/jmbi.2000.3758. PMID 10835284.

47. Krivov GG, Shapovalov MV, Dunbrack, RL; Shapovalov; Dunbrack Jr (2009). "Improved prediction of protein side-chain conformations with SCWRL4". Proteins. 77 (3): 778–795. doi:10.1002/prot.22488. PMC 2885146Freely accessible. PMID 19603484.

48. Chou KC, Zhang CT; Zhang (1995). "Prediction of protein structural classes". Crit. Rev. Biochem. Mol. Biol. 30 (4): 275–349. doi:10.3109/10409239509083488. PMID 7587280.

49. Chen C, Zhou X, Tian Y, Zou X, Cai P; Zhou; Tian; Zou; Cai (October 2006). "Predicting protein structural class with pseudo-amino acid composition and support vector machine fusion network". Anal. Biochem. 357 (1): 116–21. doi:10.1016/j.ab.2006.07.022. PMID 16920060.

50. Chen C, Tian YX, Zou XY, Cai PX, Mo JY; Tian; Zou; Cai; Mo (December 2006). "Using pseudo-amino acid composition and support vector machine to predict protein structural class". J. Theor. Biol. 243 (3): 444–8. doi:10.1016/j.jtbi.2006.06.025. PMID 16908032.

51. Lin H, Li QZ; Li (July 2007). "Using pseudo amino acid composition to predict protein structural class: approached by incorporating 400 dipeptide components". J Comput Chem. 28 (9): 1463–6. doi:10.1002/jcc.20554. PMID 17330882.

52. Xiao X, Wang P, Chou KC; Wang; Chou (October 2008). "Predicting protein structural classes with pseudo amino acid composition: an approach using geometric moments of cellular automaton image". J. Theor. Biol. 254 (3): 691–6. doi:10.1016/j.jtbi.2008.06.016. PMID 18634802.

53. Chou KC, Cai YD; Cai (September 2004). "Predicting protein structural class by functional domain composition". Biochem. Biophys. Res. Commun. 321 (4): 1007–9. doi:10.1016/j.bbrc.2004.07.059. PMID 15358128.

54. Battey JN, Kopp J, Bordoli L, Read RJ, Clarke ND, Schwede T; Kopp; Bordoli; Read; Clarke; Schwede (2007). "Automated server predictions in CASP7". Proteins. 69 (Suppl 8): 68–82. doi:10.1002/prot.21761. PMID 17894354.

55. Zhang Y, Skolnick J (2005). "The protein structure prediction problem could be solved using the current PDB library". Proc Natl Acad Sci USA. 102 (4): 1029–34. doi:10.1073/pnas.0407152101. PMC 545829Freely accessible. PMID 15653774.

56. Zemla A. (2003). "LGA — A Method for Finding 3-D Similarities in Protein Structures". Nucleic Acids Research. 31 (13): 3370–3374. doi:10.1093/nar/gkg571. PMC 168977Freely accessible. PMID 12824330.

57. Godzik A (1996). "The structural alignment between two proteins: Is there a unique answer?". Protein science : a publication of the Protein Society. 5 (7): 1325–38. doi:10.1002/pro.5560050711. PMC 2143456Freely accessible. PMID 8819165.

58. Martin ACR (1982). "Rapid Comparison of Protein Structures". Acta Crystallogr A. 38 (6): 871–873. doi:10.1107/S0567739482001806.

59. Theobald DL, Wuttke DS (2006). "Empirical Bayes hierarchical models for regularizing maximum likelihood estimation in the matrix Gaussian Procrustes problem". Proceedings of the National Academy of Sciences. 103 (49): 18521–18527. doi:10.1073/pnas.0508445103.

60. Theobald DL, Wuttke DS (2006). "THESEUS: Maximum likelihood superpositioning and analysis of macromolecular structures". Bioinformatics. 22 (17): 2171–2172. doi:10.1093/bioinformatics/btl332. PMC 2584349Freely accessible. PMID 16777907.

61. Diederichs K. (1995). "Structural superposition of proteins with unknown alignment and detection of topological similarity using a six-dimensional search algorithm". Proteins. 23 (2): 187–95. doi:10.1002/prot.340230208. PMID 8592700.

62. Maiti R, Van Domselaar GH, Zhang H, Wishart DS (2004). "SuperPose: a simple server for sophisticated structural superposition". Nucleic Acids Res. 32 (Web Server issue): W590–4. doi:10.1093/nar/gkh477. PMC 441615Freely accessible. PMID 15215457.

63. Lathrop RH. (1994). "The protein threading problem with sequence amino acid interaction preferences is NP-complete". Protein Eng. 7 (9): 1059–68. doi:10.1093/protein/7.9.1059. PMID 7831276.

64. Wang L, Jiang T (1994) On the complexity of multiple sequence alignment). "On the complexity of multiple sequence alignment". J Comput Biol. 1 (4): 337–348. doi:10.1089/cmb.1994.1.337. PMID 8790475. Check date values in: |date= (help)

65. Siew N, Elofsson A, Rychlewsk L, Fischer D (2000). "MaxSub: an automated measure for the assessment of protein structure prediction quality". Bioinformatics. 16 (9): 776–85. doi:10.1093/bioinformatics/16.9.776. PMID 11108700.

66. Poleksic A (2009). "Algorithms for optimal protein structure alignment". Bioinformatics. 25 (21): 2751–2756. doi:10.1093/bioinformatics/btp530. PMID 19734152.

67. Kolodny R, Linial N (2004). "Approximate protein structural alignment in polynomial time". PNAS. 101 (33): 12201–12206. doi:10.1073/pnas.0404383101. PMC 514457Freely accessible. PMID 15304646.

68. Martinez L, Andreani, R, Martinez, JM. (2007). "Convergent algorithms for protein structural alignment". BMC Bioinformatics. 8: 306. doi:10.1186/1471-2105-8-306. PMC 1995224Freely accessible. PMID 17714583.

69. Mount DM. (2004). Bioinformatics: Sequence and Genome Analysis 2nd ed. Cold Spring Harbor Laboratory Press: Cold Spring Harbor, NY ISBN 0879697121

70. Holm L, Sander C (1996). "Mapping the protein universe". Science. 273 (5275): 595–603. doi:10.1126/science.273.5275.595. PMID 8662544.
71. Shindyalov, I.N.; Bourne P.E. (1998). "Protein structure alignment by incremental combinatorial extension (CE) of the optimal path". Protein Engineering. 11 (9): 739–747. doi:10.1093/protein/11.9.739. PMID 9796821.
72. Prlic A, Bliven S, Rose PW, Bluhm WF, Bizon C, Godzik A, Bourne PE (2010). "Pre-calculated protein structure alignments at the RCSB PDB website". Bioinformatics (Oxford, England). 26 (23): 2983–2985. doi:10.1093/bioinformatics/btq572. PMC 3003546Freely accessible. PMID 20937596.
73. Taylor WR, Flores TP, Orengo CA (1994). "Multiple protein structure alignment". Protein Sci. 3 (10): 1858–70. doi:10.1002/pro.5560031025. PMC 2142613Freely accessible. PMID 7849601.
74. Orengo CA, Michie AD, Jones S, Jones DT, Swindells MB, Thornton JM (1997). "CATH: A hierarchical classification of protein domain structures". Structure. 5 (8): 1093–1108. doi:10.1016/S0969-2126(97)00260-8. PMID 9309224.
75. Zhang Y, Skolnick J (2005) TM–align: A protein structure alignment algorithm based on TM–score). "TM-align: A protein structure alignment algorithm based on the TM-score". Nucleic Acids Research. 33 (7): 2302–2309. doi:10.1093/nar/gki524. PMC 1084323Freely accessible. PMID 15849316. Check date values in: |date= (help)
76. Zhang Y, Skolnick J (2004) Scoring function for automated assessment of protein structure template quality). "Scoring function for automated assessment of protein structure template quality". Proteins. 57 (4): 702–710. doi:10.1002/prot.20264. PMID 15476259. Check date values in: |date= (help)
77. Stefano Angaran; Mary Ellen Bock; Claudio Garutti; Concettina Guerral (2009). "MolLoc: a web tool for the local structural alignment of molecular surfaces". Nucleic Acids Research. 37: W565–70. doi:10.1093/nar/gkp405. PMC 2703929Freely accessible. PMID 19465382.
78. Gaëlle Debret; Arnaud Martel; Philippe Cuniasse (2009). "RASMOT-3D PRO: a 3D motif search webserver". Nucleic Acids Research. 37: W459–64. doi:10.1093/nar/gkp304. PMC 2703991Freely accessible. PMID 19417073.
79. Alexandra Shulman-Peleg; Maxim Shatsky; Ruth Nussinov; Haim J. Wolfson (2008). "MultiBind and MAPPIS: webservers for multiple alignment of protein 3D-binding sites and their interactions". Nucleic Acids Research. 36: W260–4. doi:10.1093/nar/gkn185. PMC 2447750Freely accessible. PMID 18467424.
80. Alexandra Shulman-Peleg; Maxim Shatsky; Ruth Nussinov; Haim J Wolfson (2007). "Spatial chemical conservation of hot spot interactions in protein-protein complexes". BMC Biology Research. 5 (43).
81. Gabriele Ausiello; Pier Federico Gherardini; Paolo Marcatili; Anna Tramontano; Allegra Via; Manuela Helmer-Citterich (2008). "FunClust: a web server for the identification of structural motifs in a set of non-homologous protein structures". BMC Biology Research. 9.

82. Janez Konc; Dušanka Janežič (2010). "ProBiS algorithm for detection of structurally similar protein binding sites by local structural alignment" (PDF). Bioinformatics. 26 (9): 1160–1168. doi:10.1093/bioinformatics/btq100. PMC 2859123Freely accessible. PMID 20305268.
83. Hui Sun Lee; Wonpil Im (2012). "Identification of Ligand Templates using Local Structure Alignment for Structure-Based Drug Design". Journal of Chemical Information and Modeling. 52 (10): 2784–2795. doi:10.1021/ci300178e.
84. Hui Sun Lee; Wonpil Im (2013). "Ligand Binding Site Detection by Local Structure Alignment and Its Performance Complementarity". Journal of Chemical Information and Modeling. 53 (9): 2462–2470. doi:10.1021/ci4003602.
85. Barthel D., Hirst J.D., Blazewicz J., Burke E.K. and Krasnogor N. (2007). "ProCKSI: a decision support system for Protein (Structure) Comparison, Knowledge, Similarity and Information". BMC Bioinformatics. 8: 416. doi:10.1186/1471-2105-8-416. PMC 2222653Freely accessible. PMID 17963510.
86. Sippl, M.; Wiederstein, M. (2012). "Detection of spatial correlations in protein structures and molecular complexes". Structure. 20 (4): 718–728. doi:10.1016/j.str.2012.01.024. PMC 3320710Freely accessible. PMID 22483118.
87. Torarinsson E, Sawera M, Havgaard JH, Fredholm M, Gorodkin J (2006). "Torarinsson E, Sawera M, Havgaard JH, Fredholm M, Gorodkin J". Genome Res. 16 (7): 885–9. doi:10.1101/gr.5226606. PMC 1484455Freely accessible. PMID 16751343.
88. Hoksza D, Svozil D (2012). "Efficient RNA pairwise structure comparison by SETTER method". Bioinformatics. 28 (14): 1858–1864. doi:10.1093/bioinformatics/bts301. PMID 22611129.
89. Cech P, Svozil D, Hoksza D (2012). "SETTER: web server for RNA structure comparison". Nucleic Acids Research. 40 (W1): W42–W48. doi:10.1093/nar/gks560.
90. Havgaard JH, Lyngso RB, Stormo GD, Gorodkin J (2005). "Pairwise local structural alignment of RNA sequences with sequence similarity less than 40%". Bioinformatics. 21 (9): 1815–24. doi:10.1093/bioinformatics/bti279. PMID 15657094.
91. Mathews DH, Turner DH (2006). "Prediction of RNA secondary structure by free energy minimization". Curr Opin Struct Biol. 16 (3): 270–8. doi:10.1016/j.sbi.2006.05.010. PMID 16713706.
92. Levinthal C, Wodak SJ, Kahn P, Dadivanian AK (1975). "Hemoglobin Interactions in Sickle Cell Fibers: I. Theoretical Approaches to the Molecular Contacts". Proceedings of the National Academy of Sciences. 72 (4): 1330–1334. doi:10.1073/pnas.72.4.1330. PMC 432527Freely accessible. PMID 1055409.
93. Wodak SJ, Janin J (1978). "Computer Analysis of Protein-Protein Interactions". Journal of Molecular Biology. 124 (2): 323–342. doi:10.1016/0022-2836(78)90302-9. PMID 712840.
94. Wodak SJ, De Crombrugghe M, Janin J (1987). "Computer Studies of Interactions

between Macromolecules". Progress in Biophysics and Molecular Biology. 49 (1): 29–63. doi:10.1016/0079-6107(87)90008-3. PMID 3310103.

95. Katchalski-Katzir E, Shariv I, Eisenstein M, Friesem AA, Aflalo C, Vakser IA (1992). "Molecular surface recognition: determination of geometric fit between proteins and their ligands by correlation techniques". Proc. Natl. Acad. Sci. U.S.A. 89 (6): 2195–2199. doi:10.1073/pnas.89.6.2195. PMC 48623Freely accessible. PMID 1549581.
96. Gabb HA, Jackson RM, Sternberg MJ (September 1997). "Modelling protein docking using shape complementarity, electrostatics and biochemical information". J. Mol. Biol. 272 (1): 106–120. doi:10.1006/jmbi.1997.1203. PMID 9299341.
97. Strynadka NC, Eisenstein M, Katchalski-Katzir E, Shoichet BK, Kuntz ID, Abagyan R, Totrov M, Janin J, Cherfils J, Zimmerman F, Olson A, Duncan B, Rao M, Jackson R, Sternberg M, James MN (1996). "Molecular Docking Programs Successfully Predict the Binding of a Beta-lactamase Inhibitory Protein to TEM-1 Beta-Lactamase". Nature Structural & Molecular Biology. 3 (3): 233–239. doi:10.1038/nsb0396-233. PMID 8605624.
98. Gray JJ, Moughon S, Wang C, Schueler-Furman O, Kuhlman B, Rohl CA, Baker D (2003). "Protein–protein docking with simultaneous optimization of rigid-body displacement and side-chain conformations". J. Mol. Biol. 331 (1): 281–299. doi:10.1016/S0022-2836(03)00670-3. PMID 12875852.
99. Camacho CJ, Vajda S (2008). "Protein docking along smooth association pathways". Proceedings of the National Academy of Sciences. 98 (19): 10636–10641. doi:10.1073/pnas.181147798. PMC 58518Freely accessible. PMID 11517309.
100. Camacho CJ, Vajda S (2007). "In silico screening of mutational effects on enzyme-proteic inhibitor affinity: a docking-based approach". BMC structural biology. 7: 37. doi:10.1186/1472-6807-7-37. PMC 1913526Freely accessible. PMID 17559675.
101. Zhang C, Liu S, Zhu Q, Zhou Y (2005). "A knowledge-based energy function for protein–ligand, protein–protein, and protein–DNA complexes". Journal of Medicinal chemistry. 7 (48): 2325–2335. doi:10.1021/jm049314d. PMID 15801826.
102. Esmaielbeiki R, Nebel JC (2014). "Scoring docking conformations using predicted protein interfaces". BMC Bioinformatics. 15: 171. doi:10.1186/1471-2105-15-171. PMC 4057934Freely accessible. PMID 24906633.
103. Kastritis PL, Bonvin AM (May 2010). "Are scoring functions in protein–protein docking ready to predict interactomes? Clues from a novel binding affinity benchmark". J. Proteome Res. 9 (5): 2216–2225. doi:10.1021/pr9009854. PMID 20329755.
104. Rosato A, Fuentes G, Verma C (2010). "Faculty of 1000 Biology: evaluations for Kastritis PL & Bonvin AM J Proteome Res 2010 May 7 9 (5) :2216-25". Faculty of 1000 Biology.

105. Mintseris J, Wiehe K, Pierce B, Anderson R, Chen R, Janin J, Weng Z (2005). "Protein-Protein Docking Benchmark 2.0: an update". Proteins. 60 (2): 214–216. doi:10.1002/prot.20560. PMID 15981264.

106. Vreven T, Moal IH, Vangone A, Pierce BG, Kastritis PL, Torchala M, Chaleil R, Jiménez-García B, Bates PA, Fernandez-Recio J, Bonvin AM, Weng Z (September 2015). "Updates to the Integrated Protein-Protein Interaction Benchmarks: Docking Benchmark Version 5 and Affinity Benchmark Version 2". Journal of Molecular Biology. 427 (19): 3031–41. doi:10.1016/j.jmb.2015.07.016. PMC 4677049Freely accessible. PMID 26231283.

107. van Dijk M, Bonvin AM (August 2008). "A protein-DNA docking benchmark". Nucleic Acids Research. 36 (14): e88. doi:10.1093/nar/gkn386. PMC 2504314Freely accessible. PMID 18583363.

108. Barik A, C N, P M, Bahadur RP (July 2012). "A protein-RNA docking benchmark (I): nonredundant cases". Proteins. 80 (7): 1866–71. doi:10.1002/prot.24083. PMID 22488669.

109. Pérez-Cano L, Jiménez-García B, Fernández-Recio J (July 2012). "A protein-RNA docking benchmark (II): extended set from experimental and homology modeling data". Proteins. 80 (7): 1872–82. doi:10.1002/prot.24075. PMID 22488990.

110. Nithin C, Mukherjee S, Bahadur RP (November 2016). "A non-redundant protein-RNA docking benchmark version 2.0". Proteins: n/a–n/a. doi:10.1002/prot.25211. PMID 27862282.

111. Kastritis PL, Moal IH, Hwang H, Weng Z, Bates PA, Bonvin AM, Janin J (March 2011). "A structure-based benchmark for protein-protein binding affinity". Protein Science. 20 (3): 482–491. doi:10.1002/pro.580. PMC 3064828Freely accessible. PMID 21213247.

112. Janin J, Henrick K, Moult J, Eyck LT, Sternberg MJ, Vajda S, Vakser I, Wodak SJ (2003). "CAPRI: a Critical Assessment of PRedicted Interactions". Proteins. 52 (1): 2–9. doi:10.1002/prot.10381. PMID 12784359.

CHAPTER 3

Enzyme Classification

Chapter Contents

This chapter contains general information on enzyme nomenclature.

INTRODUCTION

The Enzyme Commission number (EC number) is a numerical classification scheme for enzymes, based on the chemical reactions they catalyze. As a system of enzyme nomenclature, every EC number is associated with a recommended name for the respective enzyme.

Strictly speaking, EC numbers do not specify enzymes, but enzyme-catalyzed reactions. If different enzymes (for instance from different organisms) catalyze the same reaction, then they receive the same EC number. Furthermore, through convergent evolution, completely different protein folds can catalyze an identical reaction and therefore would be assigned an identical EC number (these are called non-homologous isofunctional enzymes, or NISE). By contrast, UniProt identifiers uniquely specify a protein by its amino acid sequence.

Format of Number

Every enzyme code consists of the letters "EC" followed by four numbers separated by periods. Those numbers represent a progressively finer classification of the enzyme. Preliminary EC numbers exist and have an 'n' as part of the fourth (serial) digit (e.g. EC 3.5.1.n3).

For example, the tripeptide aminopeptidases have the code "EC 3.4.11.4", whose components indicate the following groups of enzymes:

- EC 3 enzymes are hydrolases (enzymes that use water to break up some other molecule)
- EC 3.4 are hydrolases that act on peptide bonds
- EC 3.4.11 are those hydrolases that cleave off the amino-terminal amino acid from a polypeptide
- EC 3.4.11.4 are those that cleave off the amino-terminal end from a tripeptide

Top Level Codes

Top-level EC numbers

Group	Reaction catalyzed	Typical reaction	Enzyme example(s) with trivial name
EC 1 ***Oxidoreductases***	To catalyze oxidation/reduction reactions; transfer of H and O atoms or electrons from one substance to another	$AH + B \rightarrow A + BH$ **(reduced)** $A + O \rightarrow AO$ **(oxidized)**	Dehydrogenase, oxidase
EC 2 ***Transferases***	Transfer of a functional group from one substance to another. The group may be methyl-, acyl-, amino- or phosphate group	$AB + C \rightarrow A + BC$	Transaminase, kinase
EC 3 ***Hydrolases***	Formation of two products from a substrate by hydrolysis	$AB + H_2O \rightarrow AOH + BH$	Lipase, amylase, peptidase
EC 4 ***Lyases***	Non-hydrolytic addition or removal of groups from substrates. C-C, C-N, C-O or C-S bonds may be cleaved	$RCOCOOH \rightarrow RCOH + CO_2$ or $[X\text{-}A\text{-}B\text{-}Y] \rightarrow [A{=}B + X\text{-}Y]$	Decarboxylase
EC 5 ***Isomerases***	Intramolecule rearrangement, i.e. isomerization changes within a single molecule	$ABC \rightarrow BCA$	Isomerase, mutase
EC 6 ***Ligases***	Join together two molecules by synthesis of new C-O, C-S, C-N or C-C bonds with simultaneous breakdown of ATP	$X + Y + ATP \rightarrow XY + ADP + Pi$	Synthetase

Reaction Similarity

Similarity between enzymatic reactions (EC) can be calculated by using bond changes, reaction centres or substructure metrics (EC-BLAST).

HISTORY

The enzyme nomenclature scheme was developed starting in 1955, when the International Congress of Biochemistry in Brussels set up an Enzyme Commission. The first version was published in 1961. The current sixth edition, published by the International Union of Biochemistry and Molecular Biology in 1992, contains 3196 different enzymes. Supplements 1-4 were published 1993-1999. Subsequent supplements have been published electronically, at the website of the Nomenclature Committee of the International Union of Biochemistry and Molecular Biology.

PROTEIN SUPERFAMILY

A protein superfamily is the largest grouping (clade) of proteins for which common ancestry can be inferred (see homology). Usually this common ancestry is inferred from structural alignment and mechanistic similarity, even if no sequence similarity is evident. Sequence homology can then be deduced even if not apparent (due to low sequence similarity). Superfamilies typically contain several protein families which show sequence similarity within each family. The term protein clan is commonly used for protease superfamilies based on the MEROPS protease classification system.

Identification

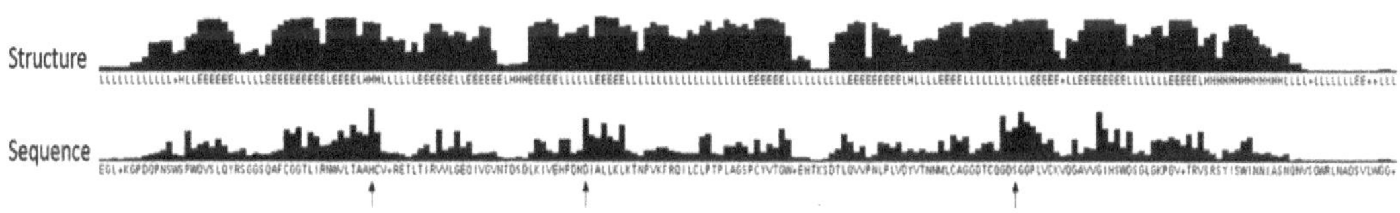

Above, secondary structural conservation of 80 members of the PA protease clan (superfamily). H indicates α-helix, E indicates β-sheet, L indicates loop. Below, sequence conservation for the same alignment. Arrows indicate catalytic triad residues. Aligned on the basis of structure by DALI.

Superfamilies of proteins are identified using a number of methods. Closely related members can be identified by different methods to those needed to group the most evolutionarily divergent members.

Sequence Similarity

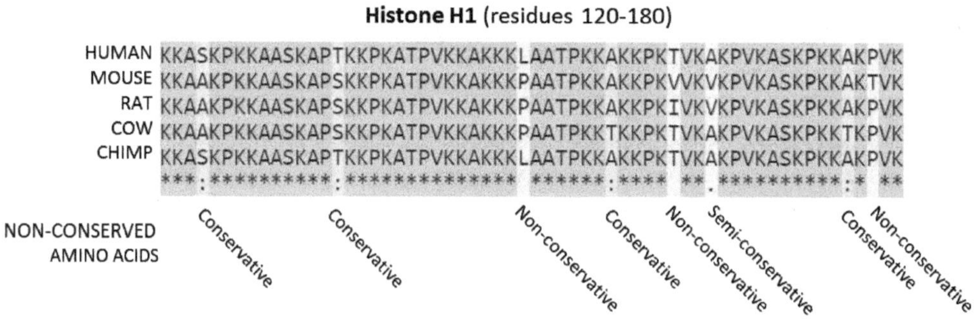

A sequence alignment of mammalian histone proteins. The similarity of the sequences implies that they evolved by gene duplication. Residues that are conserved across all sequences are highlighted in grey. Below the protein sequences is a key denoting conserved sequence (*), conservative mutations (:), semi-conservative mutations (.), and non-conservative mutations ( ).

Historically, the similarity of different amino acid sequences has been the most common method of inferring homology.Sequence similarity is considered a good predictor of relatedness, since similar sequences are more likely the result of gene duplication and divergent evolution, rather than the result of convergent evolution. Amino acid sequence is typically more conserved than DNA sequence (due to the degenerate genetic code), so is a more sensitive detection method. Since some of the amino acids have similar properties (e.g., charge, hydrophobicity, size), conservative mutations that interchange them are often neutral to function. The most conserved sequence regions of a protein often correspond to functionally important regions like catalytic sites and binding sites, since these regions are less tolerant to sequence changes.

Using sequence similarity to infer homology has several limitations. There is no minimum level of sequence similarity guaranteed to produce identical structures. Over long periods of evolution, related proteins may show no detectable sequence similarity to one another. Sequences with many insertions and deletions can also sometimes be difficult to align and so identify the homologous sequence regions. In the PA clan of proteases, for example, not a single residue is conserved through the superfamily, not even those in the catalytic triad. Conversely, the individual families that make up a superfamily are defined on the basis of their sequence alignment, for example the C04 protease family within the PA clan.

Nevertheless, sequence similarity is the most commonly used form of evidence to infer relatedness, since the number of known sequences vastly outnumbers the number of known tertiary structures. In the absence of structural information, sequence similarity constrains the limits of which proteins can be assigned to a superfamily.

Structural Similarity

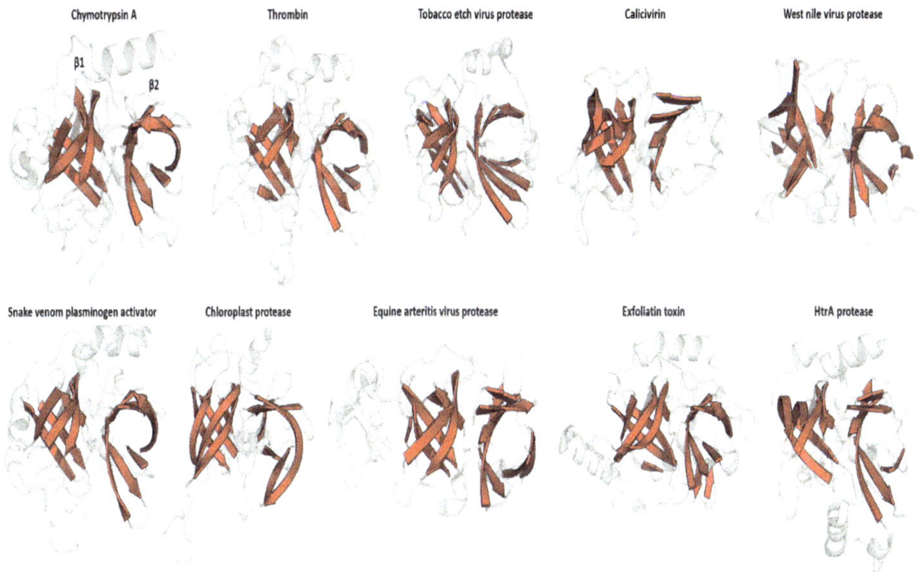

Structural homology in the PA superfamily (PA clan). The double β-barrel that characterises the superfamily is highlighted in red. Shown are representative structures from several families within the PA superfamily. Note that some proteins show partially modified structural. Chymotrypsin (1gg6), tobacco etch virus protease (1lvm), calicivirin (1wqs), west nile virus protease (1fp7), exfoliatin toxin (1exf), HtrA protease (1l1j), snake venom plasminogen activator (1bqy), chloroplast protease (4fln) and equine arteritis virus protease (1mbm).

Structurc is much more evolutionarily conserved than sequence, such that proteins with highly similar structures can have entirely different sequences. Over very long evolutionary timescales, very few residues show detectable amino acid sequence conservation, however secondary structural elements and tertiary structural motifs are highly conserved. Conformational changes of the protein structure may also be conserved, as is seen in the serpin superfamily. Consequently, protein tertiary structure can be used to detect homology between proteins even when no evidence of relatedness remains in their sequences. Structural alignment programs, such as DALI, use the 3D structure of a protein of interest to find proteins with similar folds. However, on rare occasions, related proteins may evolve to bc structurally dissimilar and relatedness can only be inferred by other methods.

Mechanistic Similarity

The catalytic mechanism of enzymes within a superfamily is typically conserved, although substrate specificity may be significantly different. Catalytic residues also tend to occur in the same order in the protein sequence. The families within the PA clan of proteases, although there has been divergent evolution of the catalytic triad residues used to perform catalysis, all members use a similar mechanism to perform covalent, nucleophilic catalysis on proteins, peptides or amino acids. However, mechanism alone is not sufficient to infer relatedness, since some catalytic mechanisms have been convergently evolved multiple times independently, and so form separate superfamilies.

Evolutionary Significance

Protein superfamilies represent the current limits of our ability to identify common ancestry. They are the largest evolutionary grouping based on direct evidence that is currently possible. They are therefore amongst the most ancient evolutionary events currently studied. Some superfamilies have members present in all kingdoms of life, indicating that the last common ancestor of that superfamily was in the last universal common ancestor of all life (LUCA).

Superfamily members may be in different species, with the ancestral protein being the form of the protein that existed in the ancestral species (orthology). Conversely, the proteins may be in the same species, but evolved from a single protein whose gene was duplicated in the genome (paralogy).

Diversification

A majority of proteins contain multiple domains. Between 66-80% of eukaryotic proteins have multiple domains while about 40-60% of prokaryotic proteins have multiple domains. Over time, many of the superfamilies of domains have mixed together. In fact, it is very rare to find "consistently isolated superfamilies". When domains do combine, the N- to C- terminal domain order (the "domain architecture") is typically well conserved. Additionally, the number of domain combinations seen in nature is small compared to the number of possibilities, suggesting that selection acts on all combinations.

Examples

α/β hydrolase superfamily - Members share an α/β sheet, containing 8 strands connected by helices, with catalytic triad residues in the same order, activities include proteases, lipases, peroxidases, esterases, epoxide hydrolases and dehalogenases.

Alkaline phosphatase superfamily - Members share an αβα sandwich structure as well as performing common promiscuous reactions by a common mechanism.

Globin superfamily - Members share an 8-helix globular globin fold.

Immunoglobulin superfamily - Members share a sandwich-like structure of two sheets of antiparallel β strands (Ig-fold), and are involved in recognition, binding, and adhesion.

PA clan - Members share a chymotrypsin-like double β-barrel fold and similar proteolysis mechanisms but sequence identity of <10%. The clan contains both cysteine and serine proteases (different nucleophiles).

Ras superfamily - Members share a common catalytic G domain of a 6-strand β sheet surrounded by 5 α-helices.

Serpin superfamily - Members share a high-energy, stressed fold which can undergo a large conformational change, which is typically used to inhibit serine and cysteine proteases by disrupting their structure.

TIM barrel superfamily - Members share a large $\alpha 8 \beta 8$ barrel structure. It is one of the most common protein folds and the monophylicity of this superfamily is still contested.

Protein Superfamily Resources

Several biological databases document protein superfamilies and protein folds, for example:

- Pfam - Protein families database of alignments and HMMs
- PROSITE - Database of protein domains, families and functional sites
- PIRSF - SuperFamily Classification System
- PASS2 - Protein Alignment as Structural Superfamilies v2
- SUPERFAMILY - Library of HMMs representing superfamilies and database of (superfamily and family) annotations for all completely sequenced organisms
- SCOP and CATH - Classifications of protein structures into superfamilies, families and domains

Similarly there are algorithms that search the PDB for proteins with structural homology to a target structure, for example:

- DALI - Structural alignment based on a distance alignment matrix method

PROTEIN FAMILY

Proteins in a family descend from a common ancestor (see homology) and typically have similar three-dimensional structures, functions, and significant sequence similarity. The most important of these is sequence similarity (usually amino acid sequence) since it is the strictest indicator of homology and therefore the clearest indicator of common ancestry. There is a fairly well developed framework for evaluating the significance of similarity between a group of sequences using sequence alignment methods. Proteins that do not share a common ancestor are very unlikely to show statistically significant sequence similarity, making sequence alignment a powerful tool for identifying the members of protein families.

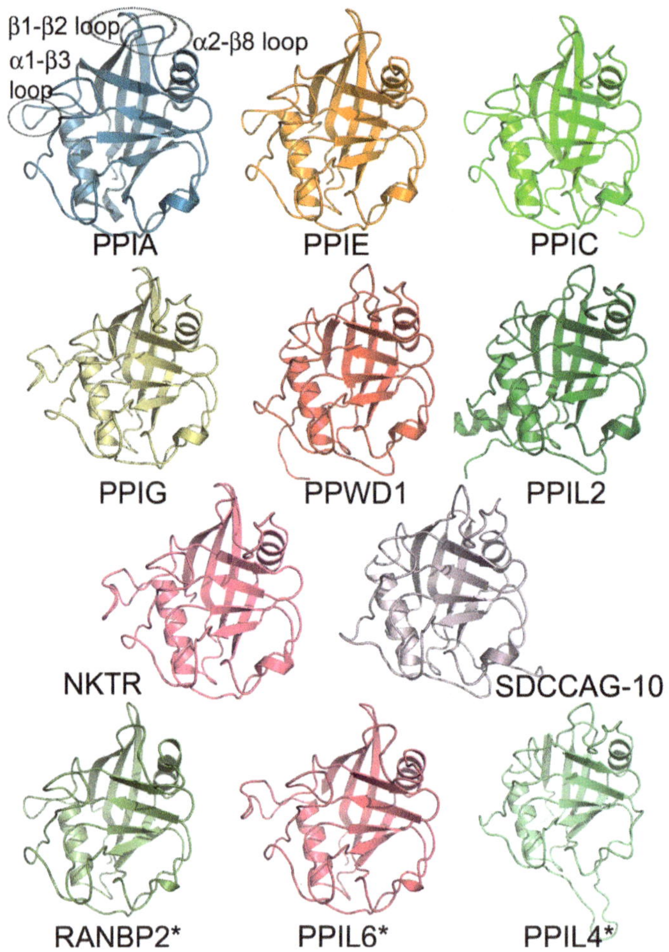

The human cyclophilin family, as represented by the structures of the isomerase domains of some of its members.

A protein family is a group of evolutionarily-related proteins. In many cases a protein family has a corresponding gene family, in which each gene encodes a corresponding protein with a 1:1 relationship. The term protein family should not be confused with family as it is used in taxonomy.

Families are sometimes grouped together into larger clades called superfamilies based on structural and mechanistic similarity, even if there is no identifiable sequence homology.

Currently, over 60,000 protein families have been defined, although ambiguity in the definition of protein family leads different researchers to wildly varying numbers.

Terminology and Usage

As with many biological terms, the use of protein family is somewhat context dependent; it may indicate large groups of proteins with the lowest possible level of detectable sequence similarity, or very narrow groups of proteins with almost identical sequence, function, and three-dimensional structure, or any kind of group in-between. To distinguish between these situations, the term protein superfamily is often used for distantly related proteins whose relatedness is not detectable by sequence similarity, but only from shared structural features. Other terms such as protein class, group, clan and sub-family have been coined over the years, but all suffer similar ambiguities of usage. A common usage is that superfamilies (structural homology) contain families (sequence homology) which contain sub-families. Hence a superfamily, such as the PA clan of proteases, has far lower sequence conservation than one of the families it contains, the C04 family. It is unlikely that an exact definition will be agreed and to it is up to the reader to discern exactly how these terms are being used in a particular context.

Protein Domains and Motifs

The concept of protein family was conceived at a time when very few protein structures or sequences were known; at that time, primarily small, single-domain proteins such as myoglobin, hemoglobin, and cytochrome c were structurally understood. Since that time, it was found that many proteins comprise multiple independent structural and functional units or domains. Due to evolutionary shuffling, different domains in a protein have evolved independently. This has led, in recent years, to a focus on families of protein domains. A number of online resources are devoted to identifying and cataloging such domains (see list of links at the end of this article).

Regions of each protein have differing functional constraints (features critical to the structure and function of the protein). For example, the active site of an enzyme requires certain amino acid residues to be precisely oriented in three dimensions. On the other hand, a protein–protein binding interface may consist of a large surface with constraints on the hydrophobicity or polarity of the amino acid residues. Functionally constrained regions of proteins evolve more slowly than unconstrained regions such as surface loops, giving rise to discernible blocks of conserved sequence when the sequences of a protein family are compared (see multiple sequence alignment). These blocks are most commonly referred to as motifs, although many other terms are used (blocks, signatures, fingerprints, etc.). Again, a large number of online resources are devoted to identifying and cataloging protein motifs (see list at end of article).

Evolution of Protein Families

According to current consensus, protein families arise in two ways. Firstly, the separation of a parent species into two genetically isolated descendent species allows a gene/protein to independently accumulate variations (mutations) in these two lineages. This results in a family of orthologous proteins, usually with conserved sequence motifs. Secondly, a gene duplication may create a second copy of a gene (termed a paralog). Because the original gene is still able to perform its function, the duplicated gene is free to diverge and may acquire new functions (by random mutation). Certain gene/protein families, especially in eukaryotes, undergo extreme expansions and contractions in the course of evolution, sometimes in concert with whole genome duplications. This expansion and contraction of protein families is one of the salient features of genome evolution, but its importance and ramifications are currently unclear.

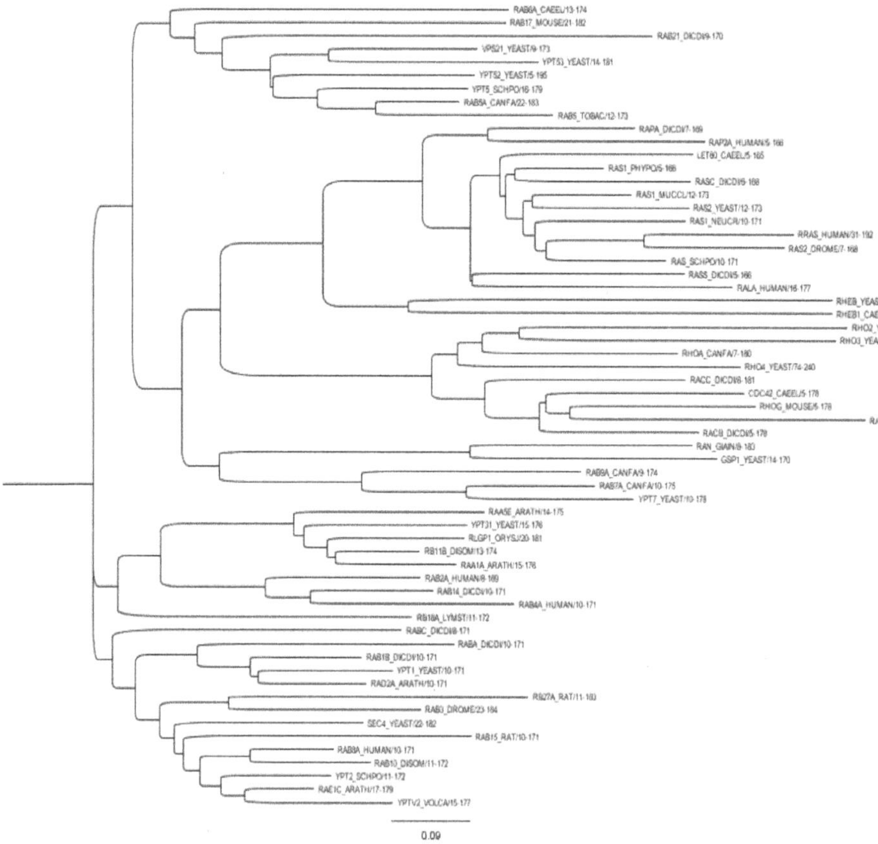

Phylogenetic tree of RAS superfamily. Tree was created using FigTree (Free online software).

Use and Importance of Protein Families

As the total number of sequenced proteins increases and interest expands in proteome analysis, there is an ongoing effort to organize proteins into families and to describe their component domains and motifs. Reliable identification of protein families is critical to phylogenetic analysis, functional annotation, and the exploration of diversity of protein function in a given phylogenetic branch. The Enzyme Function Initiative (EFI) is using protein families and superfamilies as the basis for development of a sequence/structure-based strategy for large scale functional assignment of enzymes of unknown function.

The algorithmic means for establishing protein families on a large scale are based on a notion of similarity. Most of the time the only similarity we have access to is sequence similarity.

Protein Family Resources

There are many biological databases that record examples of protein families and allow users to identify if newly identified proteins belong to a known family. Here are a few examples:

- Pfam - Protein families database of alignments and HMMs
- PROSITE - Database of protein domains, families and functional sites
- PIRSF - SuperFamily Classification System
- PASS2 - Protein Alignment as Structural Superfamilies v2 - PASS2@NCBS
- SUPERFAMILY - Library of HMMs representing superfamilies and database of (superfamily and family) annotations for all completely sequenced organisms
- SCOP and CATH - classifications of protein structures into superfamilies, families and domains

Similarly many database-searching algorithms exist, for example:

- BLAST - DNA sequence similarity search
- BLASTp - Protein sequence similarity search
- OrthoFinder: a fast, scalable and accurate method for clustering proteins into families (orthogroups)

REFERENCES

1. Holm, L; Rosenström, P (July 2010). "Dali server: conservation mapping in 3D.". Nucleic Acids Research. 38 (Web Server issue): W545–9. doi:10.1093/nar/gkq366. PMID 20457744.
2. Rawlings, ND; Barrett, AJ; Bateman, A (January 2012). "MEROPS: the database of proteolytic enzymes, their substrates and inhibitors.". Nucleic Acids Rcscarch. 40 (Database issue): D343–50. doi:10.1093/nar/gkr987. PMC 3245014. PMID 22086950.
3. "Clustal FAQ #Symbols". Clustal. Retrieved 8 December 2014.
4. Han, JH (2007). "The folding and evolution of multidomain proteins.". Nature Reviews Molecular Cell Biology. 8 (4): 319–30. doi:10.1038/nrm2144. PMID 17356578.
5. Pandit, SB, Gosar D, Abhiman S; et al. (2002). "SUPFAM--a database of potential protein superfamily relationships derived by comparing sequence-based and structure-based families: implications for structural genomics and function annotation in genomes". Nucleic Acids Research. 30 (1): 289–293. doi:10.1093/nar/30.1.289.
6. Buller, Andrew R.; Townsend, Craig A. (2013-02-19). "Intrinsic evolutionary constraints on protease structure, enzyme acylation, and the identity of the catalytic triad". Proceedings of the National Academy of Sciences. 110 (8): E653–E661. doi:10.1073/pnas.1221050110. ISSN 0027-8424. PMC 3581919. PMID 23382230.
7. Shakhnovich, BE; Deeds, E; Delisi, C; Shakhnovich, E (March 2005). "Protein structure and evolutionary history determine sequence space topology.". Genome Research. 15 (3): 385–92. doi:10.1101/gr.3133605. PMID 15741509.

8. Ranea, JA; Sillero, A; Thornton, JM; Orengo, CA (October 2006). "Protein superfamily evolution and the last universal common ancestor (LUCA).". Journal of Molecular Evolution. 63 (4): 513–25. doi:10.1007/s00239-005-0289-7. PMID 17021929.

9. Carr PD, Ollis DL (2009). "α/β hydrolase fold: an update". Protein Pept. Lett. 16 (10): 1137–48. PMID 19508187.

10. Nardini M, Dijkstra BW (December 1999). "α/β hydrolase fold enzymes: the family keeps growing". Curr. Opin. Struct. Biol. 9 (6): 732–7. doi:10.1016/S0959-440X(99)00037-8. PMID 10607665.

11. "SCOP". Retrieved 28 May 2014.

12. Mohamed, MF; Hollfelder, F (Jan 2013). "Efficient, crosswise catalytic promiscuity among enzymes that catalyze phosphoryl transfer.". Biochimica et Biophysica Acta. 1834 (1): 417–24. doi:10.1016/j.bbapap.2012.07.015. PMID 22885024.

13. Branden, Carl; Tooze, John (1999). Introduction to protein structure (2nd ed.). New York: Garland Pub. ISBN 978-0815323051.

14. Bolognesi, M; Onesti, S; Gatti, G; Coda, A; Ascenzi, P; Brunori, M (1989). "Aplysia limacina myoglobin. Crystallographic analysis at 1.6 a resolution". Journal of Molecular Biology. 205 (3): 529–44. PMID 2926816.

15. Bork P, Holm L, Sander C (September 1994). "The immunoglobulin fold. Structural classification, sequence patterns and common core". J. Mol. Biol. 242 (4): 309–20. doi:10.1006/jmbi.1994.1582. PMID 7932691.

16. Brümmendorf T, Rathjen FG (1995). "Cell adhesion molecules 1: immunoglobulin superfamily". Protein Profile. 2 (9): 963–1108. PMID 8574878.

17. Bazan, JF; Fletterick, RJ (November 1988). "Viral cysteine proteases are homologous to the trypsin-like family of serine proteases: structural and functional implications.". Proceedings of the National Academy of Sciences of the United States of America. 85 (21): 7872–6. doi:10.1073/pnas.85.21.7872. PMC 282299. PMID 3186696.

18. Vetter IR, Wittinghofer A (November 2001). "The guanine nucleotide-binding switch in three dimensions". Science. 294 (5545): 1299–304. doi:10.1126/science.1062023. PMID 11701921.

19. Silverman, G. A.; Bird, P. I.; Carrell, R. W.; Church, F. C.; Coughlin, P. B.; Gettins, P. G. W.; Irving, J. A.; Lomas, D. A.; Luke, C. J.; Moyer, R. W.; Pemberton, P. A.; Remold-O'Donnell, E.; Salvesen, G. S.; Travis, J.; Whisstock, J. C. (2 July 2001). "The Serpins Are an Expanding Superfamily of Structurally Similar but Functionally Diverse Proteins: EVOLUTION, MECHANISM OF INHIBITION, NOVEL FUNCTIONS, AND A REVISED NOMENCLATURE". Journal of Biological Chemistry. 276 (36): 33293–33296. doi:10.1074/jbc.R100016200. PMID 11435447.

20. Nagano, N; Orengo, CA; Thornton, JM (Aug 30, 2002). "One fold with many functions: the evolutionary relationships between TIM barrel families based on their sequences, structures and functions.". Journal of Molecular Biology. 321 (5): 741–65. doi:10.1016/s0022-2836(02)00649-6. PMID 12206759.

21. Farber, G (1993). "An α/β-barrel full of evolutionary trouble". Current Opinion in Structural Biology. 3 (3): 409–412. doi:10.1016/S0959-440X(05)80114-9.

CHAPTER 4

Enzymes Activity

Chapter Contents

Some substances bind the enzyme at a site other than the active site. This other site is called the allosteric site. The allosteric site allows molecules to either activate or inhibit, or turn off, enzyme activity. These molecules bind the allosteric site and change the confirmation, or shape, of the enzyme.

Molecules that turn off enzymes are called allosteric inhibitors. Allosteric inhibitors change how the active site is shaped and prevents it from binding, or attaching, to the substrate. If the enzyme can't attach to the substrate, it can't do its job! These molecules and the allosteric site they bind are like the 'off switch' for the enzyme.

INTRODUCTION

In biology, the active site is the region of an enzyme where substrate molecules bind and undergo a chemical reaction. The active site consists of residues that form temporary bonds with the substrate (binding site) and residues that catalyse a reaction of that substrate (catalytic site). The active site is usually a groove or pocket of the enzyme which can be located in a deep tunnel within the enzyme, or between the interfaces of multimeric enzymes. An active site can catalyse a reaction repeatedly as its residues are not altered at the end of the reaction (they may change during the reaction, but are regenerated by the end).

Binding Site

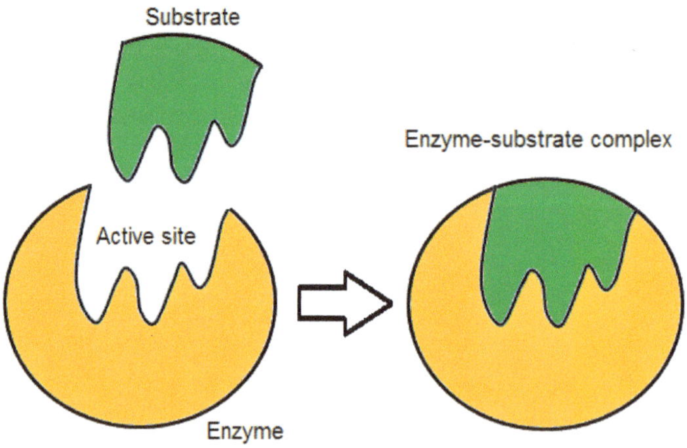

Diagram of the lock and key hypothesis

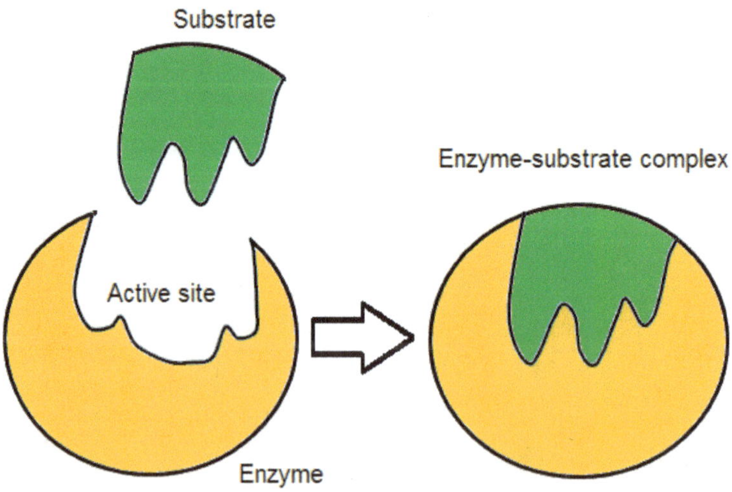

Diagram of the induced fit hypothesis.

Usually, an enzyme molecule has only one active site, and the active site fits with one specific type of substrate. An active site contains a binding site that binds the substrate and orients it for catalysis. Residues in the binding site form hydrogen bonds, hydrophobic interactions, or temporary covalent interactions (van der Waals) with the substrate to make an enzyme-substrate complex. In order to function, the active site needs to be in a specific conformation and so denaturation of the protein by high temperatures or extreme pH values will destroy its catalytic activity. A tighter fit between an active site and the substrate molecule is believed to increase efficiency of a reaction. Most enzymes have deeply buried active sites, which can be accessed by a substrate via access channels.

There are two proposed models of how enzymes fit to their specific substrate: the lock and key model and the induced fit model.

Lock and Key Hypothesis

Emil Fischer's lock and key model assumes that the active site is a perfect fit for a specific substrate and that once the substrate binds to the enzyme no further modification occurs.

Induced fit Hypothesis

Daniel Koshland's theory of enzyme-substrate binding is that the active site and the binding portion of the substrate are not exactly complementary. The induced fit model is a development of the lock-and-key model and assumes that an active site is flexible and it changes shape until the substrate is completely bound. The substrate is thought to induce a change in the shape of the active site. The hypothesis also predicts that the presence of certain residues (amino acids) in the active site will encourage the enzyme to locate the correct substrate. Conformational changes may then occur as the substrate is bound. After the products of the reaction move away from the enzyme, the active site returns to its initial shape.

CATALYTIC SITE

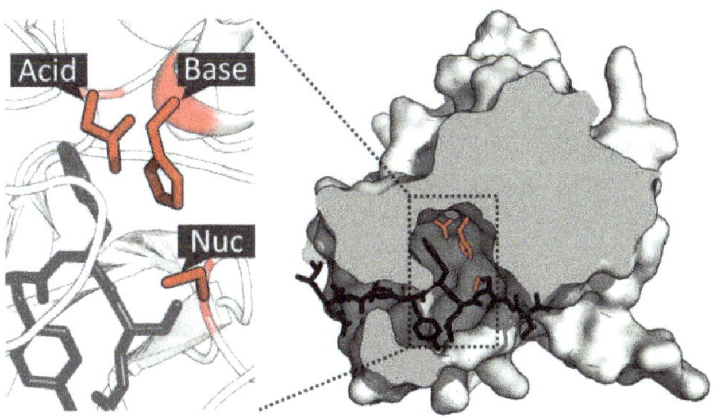

The enzyme TEV protease contains a catalytic triad of residues (red) in its catalytic site. The substrate (black) is bound by the binding site to orient it next to the triad.

Once the substrate is bound and oriented in the active site, catalysis can begin. The residues of the catalytic site are typically very close to the binding site, and some residues can have dual-roles in both binding and catalysis.

Catalytic residues of the site interact with the substrate to lower the activation energy of a reaction and so make it proceed faster. They do this by a number of different mechanisms. Firstly, they can act as donors or acceptors of protons or other groups on the substrate to facilitate the reaction. They can also form electrostatic interactions to stabilise charge buildup on the transition state or leaving group. They can also directly take part in covalent catalysis, forming an acyl-enzyme intermediate that is then resolved to regenerate the catalytic residue and release the product. In this way, the catalytic residues of the active site provide an alternative reaction mechanism with lower activation energy.

COFACTORS

Enzymes can use cofactors as 'helper molecules'. Coenzymes are one example of cofactors. Coenzymes bind to the enzyme temporarily and are released after the reaction has occurred. Metal ions are another type of cofactor.

INHIBITORS

Inhibitors disrupt the interaction between enzyme and substrate, slowing down the rate of a reaction. There are different types of inhibitor, including both reversible and irreversible forms. Reversible inhibitors can be competitive or non-competitive. Competitive reversible inhibitors have a similar shape to the substrate and bind to the enzyme's active site temporarily, blocking entry of the actual substrate into the active site. Non-competitive reversible inhibitors bind to the enzyme however not in the active site. Despite not interacting with the active site, non-competitive inhibitors do reduce the rate of the reaction because they cause the enzyme to change shape. Irreversible inhibitors bind permanently to the enzyme, blocking access to active sites and therefore reducing the rate of the reaction.

	Example	Binds active site?	Reduces rate of reaction?
Competitive reversible inhibitor	HIV protease inhibitors	Yes	Yes
Non-competitive reversible inhibitor	Heavy metals such as lead and mercury	No	Yes
Irreversible inhibitor	Cyanide	Yes	Yes

IN DRUG DISCOVERY

Identification of active sites is crucial in the process of drug discovery. The 3-D structure of the enzyme is analysed to identify active sites and design drugs which can fit into them. Proteolytic enzymes are targets for some drugs, such as protease inhibitors, which include drugs against AIDS and hypertension. These protease inhibitors bind to an enzyme's active site and block interaction with natural substrates. An important factor in drug design is the strength of binding between the active site and an enzyme inhibitor.

Active sites can be mapped to aid design of new drugs such as enzyme inhibitors. This involves description of the size of an active site and the number and properties of sub-sites, such as details of the binding interaction. Modern database technology called CPASS (Comparison of Protein Active Site Structures) however allows us to compare active sites in more detail and to look at structural similarity using software.

ALLOSTERIC SITES

An allosteric site is a site on an enzyme, unrelated to its active site, which can bind an effector molecule. This interaction is another mechanism of enzyme regulation. Allosteric modification usually happens in proteins with more than one subunit. Allosteric interactions are often present in metabolic pathways and are beneficial in that they allow one step of a reaction to regulate another step. They allow an enzyme to have a range of molecular interactions, other than the highly specific active site.

BINDING SITE

In biochemistry, a binding site is a region on a protein or piece of DNA or RNA to which ligands (specific molecules and/or ions) may form a chemical bond. An equilibrium exists between unbound ligands and bound ligands.

Saturation is the fraction of total binding sites that are occupied at any given time. When more than one type of ligand can bind to a binding site, competition ensues.

Binding sites also exhibit chemical specificity, a measure of the types of ligands that will bond, and affinity, which is a measure of the strength of the chemical bond.

Binding sites are often an important component of the functional characterization of biomolecules. For example, the characterization of the active site of a substrate to an enzyme is essential to model the reaction mechanism responsible for the chemical change from substrate to product.

Binding sites on proteins can sometimes recognize other proteins. When a binding site of one protein identifies with another protein's surface, a non-covalent bond is formed between the two polypeptide (peptide) chains and a combined new protein is formed.

A more specific type of binding site is the transcription factor binding site present on DNA. Binding sites also exist on antibodies as specifically coded regions that bind antigens based upon their structure. Several supervised Machine learning models and applications were suggested to identify the binding sites.

CATALYTIC TRIAD

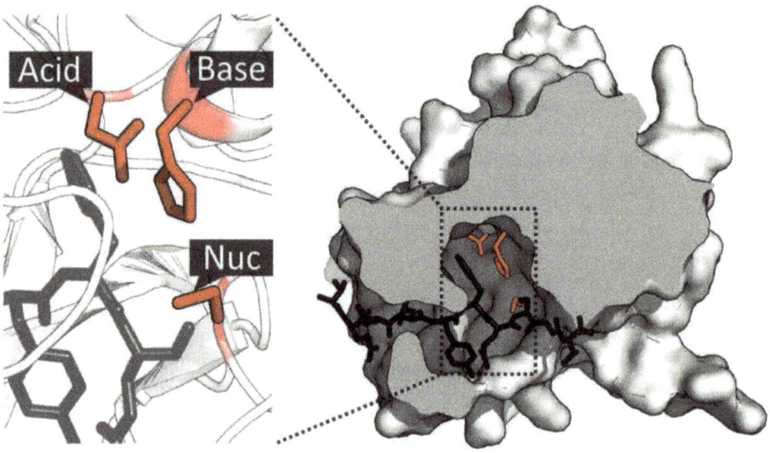

The enzyme TEV protease[a] contains an example of a catalytic triad of residues (red) in its active site. The triad consists of an aspartate (Acid), histidine (Base) and serine (Nucleophile). The substrate (black) is bound by the binding site to orient it next to the triad.

A catalytic triad refers to the three amino acid residues that function together at the centre of the active site of some hydrolase and transferase enzymes (e.g. proteases, amidases, esterases, acylases, lipases and β-lactamases). An Acid-Base-Nucleophile triad is a common motif for generating a nucleophilic residue for covalent catalysis. The residues form a charge-relay network to polarise and activate the nucleophile, which attacks the substrate, forming a covalent intermediate which is then hydrolysed to regenerate free enzyme and release the product. The nucleophile is most commonly a serine or cysteine amino acid, but occasionally threonine. Because enzymes fold into complex three-dimensional structures, the residues of a catalytic triad can be far from each other along the amino-acid sequence (primary structure), however, they are brought close together in the final fold.

As well as divergent evolution of function (and even the triad's nucleophile), catalytic triads show some of the best examples of convergent evolution. Chemical constraints on catalysis have led to the same catalytic solution independently evolving in at least 23 separate superfamilies. Their mechanism of action is consequently one of the best studied in biochemistry.

The enzymes trypsin and chymotrypsin were first purified in the 1930s. A serine in each of trypsin and chymotrypsin was identified as the catalytic nucleophile (by diisopropyl fluorophosphate modification) in the 1950s. The structure of chymotrypsin was solved by X-ray crystallography in the 1960s, showing the orientation of the catalytic triad in the active site. Other proteases were sequenced and aligned to reveal a family of related proteases, now called the S1 family. Simultaneously, the structures of the evolutionarily unrelated papain and subtilisin proteases were found to contain analogous triads. The 'charge-relay' mechanism for the activation of the nucleophile by the other triad members was proposed in the late 1960s. As more protease structures were solved by X-ray crystallography in the 1970s and 80s, homologous (such as TEV protease) and analogous (such as papain) triads were found. The MEROPS classification system in the 1990s and 2000s began classing proteases into structurally related enzyme superfamilies and so acts as a database of the convergent evolution of triads in over 20 superfamilies. Understanding how chemical constraints on evolution led to the convergence of so many enzyme families on the same triad geometries has developed in the 2010s. Of particular contention during the 1990s and early 2000s was the contribution of low-barrier hydrogen bonding to catalysis; however, current thinking is that ordinary hydrogen bonding is sufficient to explain the mechanism. The massive body of work on the charge-relay, covalent catalysis used by catalytic triads has led to the mechanism being the best characterised in all of biochemistry.

Function

Enzymes that contain an catalytic triad use it for one of two reaction types: either to split a substrate (hydrolases) or to transfer one portion of a substrate over to a second substrate (transferases). Triads are an inter-dependent set of residues in the active site of an enzyme and act in concert with other residues (e.g. binding site and oxyanion hole) to achieve nucleophilic catalysis. These triad residues act together to make the nucleophile member highly reactive, generating a covalent intermediate with the substrate that is then resolved to complete catalysis.

Mechanism

Catalytic triads perform covalent catalysis using a residue as a nucleophile. The reactivity of the nucleophilic residue is increased by the functional groups of the other triad members. The nucleophile is polarised and oriented by the base, which is itself bound and stabilised by the acid.

Catalysis is performed in two stages. First, the activated nucleophile attacks the carbonyl carbon and forces the carbonyl oxygen to accept an electron, leading to a tetrahedral intermediate. The build-up of negative charge on this intermediate is typically stabilized by an oxanion hole within the active site. The intermediate then collapses back to a carbonyl, ejecting the first half of the substrate, but leaving the second half still covalently bound to the enzyme as an acyl-enzyme intermediate. The ejection of this first leaving group is often aided by donation of a proton by the base.

The second stage of catalysis is the resolution of the acyl-enzyme intermediate by the attack of a second substrate. If this substrate is water then the result is hydrolysis; if it is an organic molecule then the result is transfer of that molecule onto the first substrate. Attack by this second substrate forms a new tetrahedral intermediate, which resolves by ejecting the enzyme's nucleophile, releasing the second product and regenerating free enzyme.

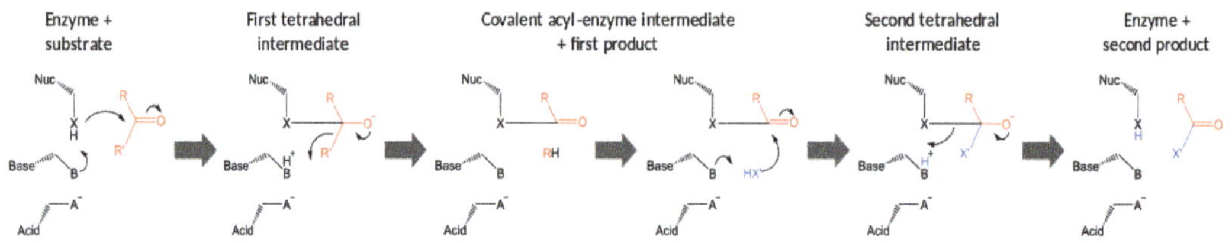

General reaction mechanism of catalysed by a catalytic triad (black): nucleophilic substitution at a carbonyl substrate (red) by a second substrate (blue). First, the enzyme's nucleophile (X) attacks the carbonyl to form a covalently linked acyl-enzyme intermediate. This intermediate is then attacked by the second substrate's nucleophile (X'). If the second nucleophile is the hydroxyl of water, the result is hydrolysis, otherwise the result is group transfer of X'.

Identity of Triad Members

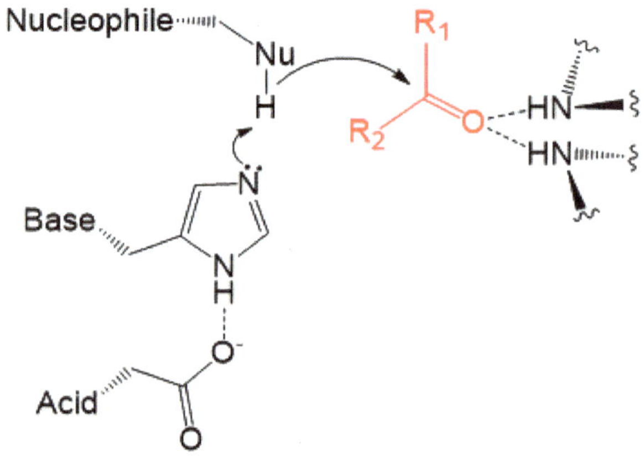

A catalytic triad charge-relay system as commonly found in proteases. The acid residue (commonly glutamate or aspartate) aligns and polarises the base (usually histidine) which activates the nucleophile (often serine or cysteine, occasionally threonine). The triad reduces the pK_a of the nucleophilic residue which then attacks the substrate. An oxyanion hole of positively charged usually backbone amides (occasionally side-chains) stabilise charge build-up on the substrate transition state.

Nucleophile

The side-chain of the nucleophilic residue performs covalent catalysis on the substrate. The lone pair of electrons present on the oxygen or sulphur attacks the electropositive carbonyl carbon. The 20 naturally occurring biological amino acids do not contain any sufficiently nucleophilic functional groups for many difficult catalytic reactions. Embedding the nucleophile in a triad increases its reactivity for efficient catalysis. The most commonly used nucleophiles are the hydroxyl (OH) of serine and the thiol/thiolate ion (SH/S−) of cysteine. Alternatively, threonine proteases use the secondary hydroxyl of threonine, however due to steric hindrance of the side chain's extra methyl group such proteases use their N-terminal amide as the base, rather than a separate amino acid.

Use of oxygen or sulphur as the nucleophilic atom causes minor differences in catalysis. Compared to oxygen, sulphur's extra d orbital makes it larger (by 0.4 Å) and softer, allows it to form longer bonds (dC-X and dX-H by 1.3-fold), and gives it a lower pKa (by 5 units). Serine is therefore more dependent than cysteine on optimal orientation of the acid-base triad members to reduce its pKain order to achieve concerted deprotonation with catalysis. The low pKa of cysteine works to its disadvantage in the resolution of the first tetrahedral intermediate as unproductive reversal of the original nucleophilic attack is the more favourable breakdown product.The triad base is therefore preferentially oriented to protonate the leaving group amide to ensure that it is ejected to leave the enzyme sulphur covalently bound to the substrate N-terminus. Finally, resolution of the acyl-enzyme (to release the substrate C-terminus) requires serine to be re-protonated whereas cysteine can leave as S−. Sterically, the sulphur of cysteine also forms longer bonds and has a bulkier Van der Waals radius and if mutated to serine can be trapped in unproductive orientations in the active site.

Base

Since no natural amino acids are strongly nucleophilic, the base in a catalytic triad polarises and deprotonates the nucleophile to increase its reactivity. Additionally, it protonates the first product to aid leaving group departure.

The base is most commonly histidine since its pKa allows for effective base catalysis, hydrogen bonding to the acid residue, and deprotonation of the nucleophile residue. β-lactamases such as TEM-1 use a lysine residue as the base. Because lysine's pKa is so high (pKa=11), a glutamate and several other residues act as the acid to stabilise its deprotonated state during the catalytic cycle. Threonine proteases use their N-terminal amide as the base, since steric crowding by the catalytic threonine's methyl prevents other residues from being close enough.

Acid

The acidic triad member forms a hydrogen bond with the basic residue. This aligns the basic residue by restricting its side-chain rotation, and polarises it by stabilising its positive charge.Two amino acids have acidic side chains at physiological pH (aspartate or glutamate) and so are the most commonly used for this triad member. Cytomegalovirus protease[b] uses a pair of histidines, one as the base, as usual, and one as the acid. The second histidine is not as effective an acid as the more common aspartate or glutamate, leading to a lower catalytic efficiency. In some enzymes, the acid member of the triad is less necessary and some act only as a dyad. For example, papain[c] uses asparagine as its third triad member which orients the histidine base but does not act as an acid. Similarly, hepatitis A virus protease[d] contains an ordered water in the position where an acid residue should be.

Examples of Triads

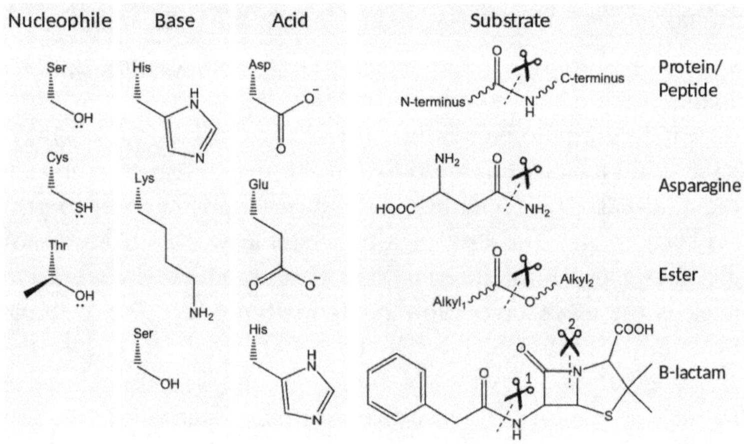

The range of amino acid residues used in different combinations in different enzymes to make up a catalytic triad for hydrolysis. On the left are the nucleophile, base and acid triad members. On the right are different substrates with the cleaved bond indicated by a pair of scissors. Two different bonds in beta-lactams can be cleaved (1 by penicillin acylase and 2 by beta-lactamase)

Ser-His-Asp

The Serine-Histidine-Aspartate motif is one of the most thoroughly characterised catalytic motifs in biochemistry. The triad is exemplified by chymotrypsin,[e] a model serine protease from the PA superfamily which uses its triad to hydrolyse protein backbones. The aspartate is hydrogen bonded to the histidine, increasing the pKa of its imidazole nitrogen from 7 to around 12. This allows the histidine to act as a powerful general base and to activate the serine nucleophile. It also has an oxyanion hole consisting of several backbone amides which stabilises charge build-up on intermediates. The histidine base aids the first leaving group by donating a proton, and also activates the hydrolytic water substrate by abstracting a proton as the remaining OH− attacks the acyl-enzyme intermediate.

The same triad has also convergently evolved in α/β hydrolases such as some lipases and esterases, however orientation of the triad members is reversed. Additionally, brain acetyl hydrolase (which has the same fold as a small G-protein) has also been found to have this triad. The equivalent Serine-Histidine-Glutamate triad is used in acetylcholinesterase.

Cys-His-Asp

The second most studied triad is the Cysteine-Histidine-Aspartate motif. Several families of cysteine proteases use this triad set, for example TEV protease[a] and papain.[c] The triad acts similarly to serine protease triads, with a few notable differences. Due to cysteine's low pKa, the importance of the Asp to catalysis varies and several cysteine proteases are effectively Cys-His dyads (e.g. hepatitis A virus protease), whilst in others the cysteine is already deprotonated before catalysis begins (e.g. papain). This triad is also used by some amidases, such as N-glycanase to hydrolyse non-peptide C-N bonds.

Ser-His-His

The triad of cytomegalovirus protease[b] uses histidine as both the acid and base triad members. Removing the acid histidine results in only a 10-fold activity loss (compared to >10,000-fold when aspartate is removed from chymotrypsin). This triad has been interpreted as a possible way of generating a less active enzyme to control cleavage rate.

Ser-Glu-Asp

An unusual triad is found in seldolisin proteases.[f] The low pKa of the glutamate carboxylate group means that it only acts as a base in the triad at very low pH. The triad is hypothesised to be an adaptation to specific environments like acidic hot springs (e.g. kumamolysin) or cell lysosome (e.g. tripeptidyl peptidase).

Thr-Nter, Ser-Nter and Cys-Nter

Threonine proteases, such as the proteasome protease subunit[g] and ornithine acyltransferases[h] use the secondary hydroxyl of threonine in an manner analogous to the use of the serine primary hydroxyl. However, due to the steric interference of the extra methyl group of threonine, the base member of the triad is the N-terminal amide which polarises an ordered water which, in turn, deprotonates the catalytic hydroxyl to increase its reactivity. Similarly, there exist equivalent 'serine only' and 'cysteine only' configurations such as penicillin acylase G[i] and penicillin acylase V[j] which are evolutionarily related to the proteasome proteases. Again, these use their N-terminal amide as a base.

Ser-cisSer-Lys

This unusual triad occurs only in one superfamily of amidases. In this case, the lysine acts to polarise the middle serine. The middle serine then forms two strong hydrogen bonds to the nucleophilic serine to activate it (one with the side chain hydroxyl and the other with the backbone amide). The middle serine is held in an unusual cis orientation to facilitate precise contacts with the other two triad residues. The triad is further unusual in that the lysine and cis-serine both act as the base in activating the catalytic serine, but the same lysine also performs the role of the acid member as well as making key structural contacts.

Divergent Evolution

The sophistication of the active site network causes residues involved in catalysis (and residues in contact with these) to be highly evolutionarily conserved. However, there are examples of divergent evolution in catalytic triads, both in the reaction catalysed, and the residues used in catalysis. The triad remains the core of the active site, but it is evolutionarily adapted to serve different functions.

Reaction Changes

Catalytic triads perform covalent catalysis via an acyl-enzyme intermediate. If this intermediate is resolved by water, the result is hydrolysis of the substrate. However, if the intermediate is resolved by attack by a second substrate, then the enzyme acts as a transferase. For example, attack by an acyl group results in an acyltransferase reaction. Several families of transferase enzymes have evolved from hydrolases by adaptation to exclude water and favour attack of a second substrate.

Additionally, an alternative transferase mechanism has been evolved by amidophosphoribosyltransferases, which has two active sites.[k] In the first active site, a cysteine triad hydrolyses a glutamine substrate to release free ammonia. The ammonia then diffuses though an internal tunnel in the enzyme to the second active site, where it is transferred to a second substrate.

Nucleophile Changes

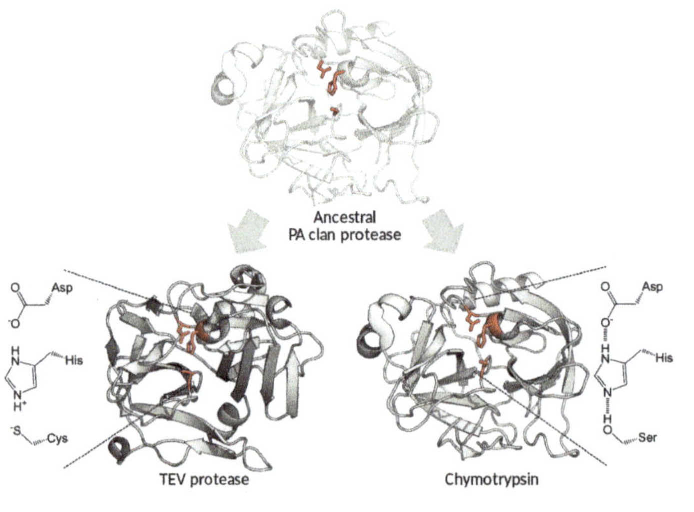

Divergent evolution of PA clan proteases to use different nucleophiles in their catalytic triad. Shown are the serine triad of chymotrypsin[e] and the cysteine triad of TEV protease.[a]

Divergent evolution of active site residues is slow, due to strong chemical constraints. Nevertheless, some protease superfamilies have evolved from one nucleophile to another. This can be inferred when a superfamily (with the same fold) contains families that use different nucleophiles. Such nucleophile switches have occurred several times during evolutionary history, however the mechanisms by which this happen are still unclear.

Within protease superfamilies that contain a mixture of nucleophiles (e.g. the PA clan), families are designated by their catalytic nucleophile (C=cysteine proteases, S=serine proteases).

<table>
<tr><td colspan="3">Superfamilies containing a mixture of families that use different nucleophiles [hide]</td></tr>
<tr><th>Superfamily</th><th>Families</th><th>Examples</th></tr>
<tr><td rowspan="2">PA clan</td><td>C3, C4, C24, C30, C37, C62, C74, C99</td><td>TEV protease (Tobacco etch virus)</td></tr>
<tr><td>S1, S3, S6, S7, S29, S30, S31, S32, S39, S46, S55, S64, S65, S75</td><td>Chymotrypsin (mammals, e.g. Bos taurus)</td></tr>
<tr><td rowspan="3">PB clan</td><td>C44, C45, C59, C69, C89, C95</td><td>Amidophosphoribosyltransferase precursor (Homo sapiens)</td></tr>
<tr><td>S45, S63</td><td>Penicillin G acylase precursor (Escherichia coli)</td></tr>
<tr><td>T1, T2, T3, T6</td><td>Archaean proteasome, beta component (Thermoplasma acidophilum)</td></tr>
<tr><td rowspan="2">PC clan</td><td>C26, C56</td><td>Gamma-glutamyl hydrolase (Rattus norvegicus)</td></tr>
<tr><td>S51</td><td>Dipeptidase E (Escherichia coli)</td></tr>
<tr><td rowspan="2">PD clan</td><td>C46</td><td>Hedgehog protein (Drosophila melanogaster)</td></tr>
<tr><td>N9, N10, N11</td><td>Intein-containing V-type proton ATPase catalytic subunit A (Saccharomyces cerevisiae)</td></tr>
<tr><td rowspan="2">PE clan</td><td>P1</td><td>DmpA aminopeptidase (Ochrobactrum anthropi)</td></tr>
<tr><td>T5</td><td>Ornithine acetyltransferase precursor (Saccharomyces cerevisiae)</td></tr>
</table>

Convergent Evolution

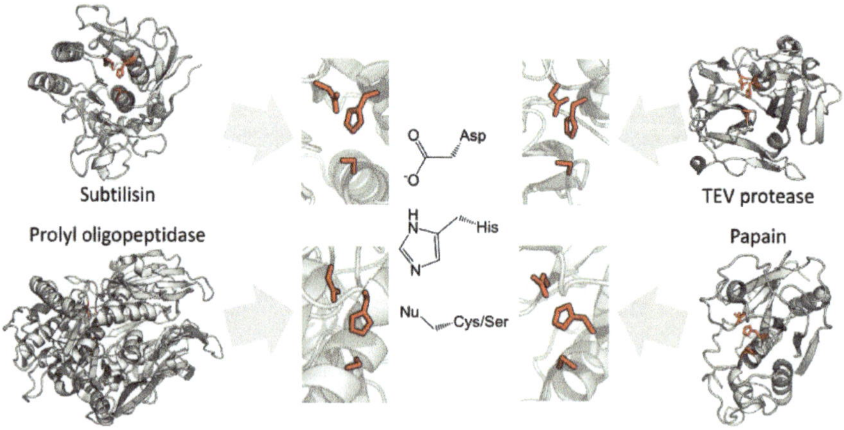

Evolutionary convergence of serine and cysteine protease towards the same catalytic triads organisation of acid-base-nucleophile in different protease superfamilies. Shown are the triads of subtilisin,[l] prolyl oligopeptidase,[m] TEV protease,[a] and papain.[c]

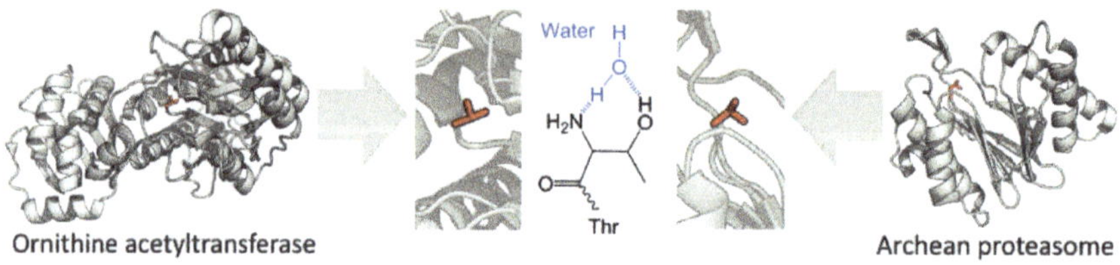

Evolutionary convergence of threonine proteases towards the same *N*-terminal active site organisation. Shown are the catalytic threonine of the proteasome[g] and ornithine acetyltransferase.[h]

The enzymology of proteases provides some of the clearest known examples of convergent evolution. The same geometric arrangement of triad residues occurs in over 20 separate enzyme superfamilies. Each of these superfamilies is the result of convergent evolution for the same triad arrangement within a different structural fold. This is because there are limited productive ways to arrange three triad residues, the enzyme backbone and the substrate. These examples reflect the intrinsic chemical and physical constraints on enzymes, leading evolution to repeatedly and independently converge on equivalent solutions.

Cysteine and Serine Hydrolases

The same triad geometries been converged upon by serine proteases such as the chymotrypsin[e] and subtilisin superfamilies. Similar convergent evolution has occurred with cysteine proteases such as viral C3 protease and papain[c] superfamilies. These triads have converged to almost the same arrangement due to the mechanistic similarities in cysteine and serine proteolysis mechanisms.

Threonine Proteases

Threonine proteases use the amino acid threonine as their catalytic nucleophile. Unlike cysteine and serine, threonine is a secondary hydroxyl (i.e. has a methyl group). This methyl group greatly restricts the possible orientations of triad and substrate as the methyl clashes with either the enzyme backbone or histidine base. Consequently, the catalytic residue of a threonine protease is located at it N-terminus.

Two evolutionarily independent enzyme superfamilies with different protein folds are known to use the N-terminal residue as a nucleophile: Superfamily PB (proteasomes using the Ntn fold) and Superfamily PE (acetyltransferases using the DOM fold) This commonality of active site structure in completely different protein folds indicates that the active site evolved convergently in those superfamilies.

OXYANION HOLE

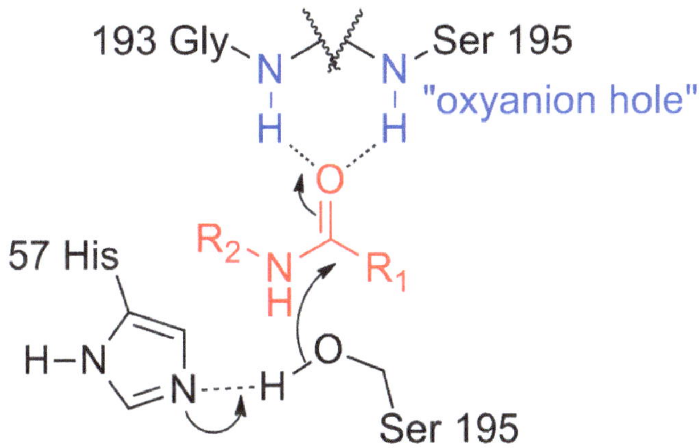

Oxyanion hole of a serine protease (black) stabilises negative charge build-up on the transition state of the substrate (red) using hydrogen bonds from enzyme's backbone amides (blue).

An oxyanion hole is a pocket in the active site of an enzyme that stabilizes transition state negative charge on a deprotonated oxygen or alkoxide. The pocket typically consists of backbone amides or positively charged residues. Stabilising the transition state lowers the activation energy necessary for the reaction, and so promotes catalysis. For example, proteases such as chymotrypsin contain an oxianion hole to stabilise the tetrahedral intermediate anion formed during proteolysis. Additionally, it may allow for insertion or positioning of a substrate, which would suffer from steric hindrance if it could not occupy the hole (such as BPG in hemoglobin). Enzymes that catalyse multi-step reactions can have multiple oxyanion holes that stabilise different transition states in the reaction.

Enzyme promiscuity

Enzyme promiscuity is the ability of an enzyme to catalyse a fortuitous side reaction in addition to its main reaction. Although enzymes are remarkably specific catalysts, they can often perform side reactions in addition to their main, native catalytic activity. These promiscuous activities are usually slow relative to the main activity and are under neutral selection. Despite ordinarily being physiologically irrelevant, under new selective pressures these activities may confer a fitness benefit therefore prompting the evolution of the formerly promiscuous activity to become the new main activity. An example of this is the atrazine chlorohydrolase (atzA encoded) from Pseudomonas sp. ADP which evolved from melamine deaminase (triA encoded), which has very small promiscuous activity towards atrazine, a man-made chemical.

Enzymes are evolved to catalyse a particular reaction on a particular substrate with a high catalytic efficiency (kcat/KM, cf. Michaelis–Menten kinetics). However, in addition to this main activity, they possess other activities that are generally several orders of magnitude lower, and that are not a result of evolutionary selection and therefore do not partake in the physiology of the organism.[nb 1] This phenomenon allows new functions to be gained as the promiscuous activity could confer a fitness benefit under a new selective pressure leading to its duplication and selection as a new main activity.

Enzyme Evolution

Duplication and Divergence

Several theoretical models exist to predict the order of duplication and specialisation events, but the actual process is more intertwined and fuzzy (§ Reconstructed enzymes below). On one hand, gene amplification results in an increase in enzyme concentration, and potentially freedom from a restrictive regulation, therefore increasing the reaction rate (v) of the promiscuous activity of the enzyme making its effects more pronounced physiologically ("gene dosage effect"). On the other, enzymes may evolve an increased secondary activity with little loss to the primary activity ("robustness") with little adaptive conflict (§ Robustness and plasticity below).

Robustness and Plasticity

A study of three distinct hydrolases (human serum paraoxonase (PON1), pseudomonad phosphotriesterase (PTE) and human carbonic anhydrase II (CAII)) has shown the main activity is "robust" towards change, whereas the promiscuous activities are more "plastic". Specifically, selecting for an activity that is not the main activity (via directed evolution), does not initially diminish the main activity (hence its robustness), but greatly affects the non-selected activities (hence their plasticity).

The phosphotriesterase (PTE) from Pseudomonas diminuta was evolved to become an arylesterase (P–O to C–O hydrolase) in eighteen rounds gaining a 109 shift in specificity (ratio of KM), however most of the change occurred in the initial rounds, where the unselected vestigial PTE activity was retained and the evolved arylesterase activity grew, while in the latter rounds there was a little trade-off for the loss of the vestigial PTE activity in favour of the arylesterase activity.

This means firstly that a specialist enzyme (monofunctional) when evolved goes through a generalist stage (multifunctional), before becoming a specialist again—presumably after gene duplication according to the IAD model—and secondly that promiscuous activities are more plastic than the main activity.

Reconstructed Enzymes

The most recent and most clear cut example of enzyme evolution is the rise of bioremediating enzymes in the past 60 years. Due to the very low number of amino acid changes, these provide an excellent model to investigate enzyme evolution in nature. However, using extant enzymes to determine how the family of enzymes evolved has the drawback that the newly evolved enzyme is compared to paralogues without knowing the true identity of the ancestor before the two genes divereged. This issue can be resolved thanks to ancestral reconstruction. First proposed in 1963 by Linus Pauling and Emile Zuckerkandl, ancestral reconstruction is the inference and synthesis of a gene from the ancestral form of a group of genes, which has had a recent revival thanks to improved inference techniques and low-cost artificial gene synthesis, resulting in several ancestral enzymes—dubbed "stemzymes" by some—to be studied.

Evidence gained from reconstructed enzyme suggests that the order of the events where the novel activity is improved and the gene is duplication is not clear cut, unlike what the theoretical models of gene evolution suggest.

One study showed that the ancestral gene of the immune defence protease family in mammals had a broader specificity and a higher catalytic efficiency than the contemporary family of paralogues, whereas another study showed that the ancestral steroid receptor of vertebrates was an oestrogen receptor with slight substrate ambiguity for other hormones—indicating that these probably were not synthesised at the time.

This variability in ancestral specificity has not only been observed between different genes, but also within the same gene family. In light of the large number of paralogous fungal α-glucosidase genes with a number of specific maltose-like (maltose, turanose, maltotriose, maltulose and sucrose) and isomaltose-like (isomaltose and palatinose) substrates, a study reconstructed all key ancestors and found that the last common ancestor of the paralogues was mainly active on maltose-like substrates with only trace activity for isomaltose-like sugars, despite leading to a lineage of iso-maltose glucosidases and a lineage that further split into maltose glucosidases and iso-maltose glucosidases. Antithetically, the ancestor before the latter split had a more pronounced isomaltose-like glucosidase activity.

Primordial Metabolism

Roy Jensen in 1976 theorised that primordial enzymes had to be highly promiscuous in order for metabolic networks to assemble in a patchwork fashion (hence its name, the patchwork model). This primordial catalytic versatility was later lost in favour of highly catalytic specialised orthologous enzymes. As a consequence, many central-metabolic enzymes have structural homologues that diverged before the last universal common ancestor.

Distribution

Promiscuity is however not only a primordial trait, in fact it is very widespread property in modern genomes. A series of experiments have been conducted to assess the distribution of promiscuous enzyme activities in E. coli. In E. coli 21 out of 104 single-gene knockouts tested (from the Keio collection) could be rescued by overexpressing a noncognate E. coli protein (using a pooled set of plasmids of the ASKA collection). The mechanisms by which the noncognate ORF could rescue the knockout can be grouped into eight categories: isozyme overexpression (homologues), substrate ambiguity, transport ambiguity (scavenging), catalytic promiscuity, metabolic flux maintenance (including overexpression of the large component of a synthase in the absence of the amine transferase subunit), pathway bypass, regulatory effects and unknown mechanisms. Similarly, overexpressing the ORF collection allowed E. coli to gain over an order of magnitude in resistance in 86 out 237 toxic environment.

Homology

Homologues are sometimes known to display promiscuity towards each other's main reactions. This crosswise promiscuity has been most studied with members of the alkaline phosphatase superfamily, which catalyse hydrolytic reaction on the sulfate, phosphonate, monophosphate, diphosphate or triphosphate ester bond of several compounds. Despite the divergence the homologues have a varying degree of reciprocal promiscuity: the differences in promiscuity are due to mechanisms involved, particularly the intermediate required.

Degree of Promiscuity

Enzymes are generally in a state that is not only a compromise between stability and catalytic efficiency, but also for specificity and evolvability, the latter two dictating whether an enzyme is a generalist (highly evolvable due to large promiscuity, but low main activity) or a specialist (high main activity, poorly evolvable due to low promiscuity). Examples of these are enzymes for primary and secondary metabolism in plants (§ Plant secondary metabolism below). Other factors can come into play, for example the glycerophosphodiesterase (gpdQ) from Enterobacter aerogenes shows different values for its promiscuous activities depending on the two metal ions it binds, which is dictated by ion availability. In some cases promiscuity can be increased by relaxing the specificity of the active site by enlarging it with a single mutation as was the case of a D297G mutant of the E. coli L-Ala-D/L-Glu epimerase (ycjG) and E323G mutant of a pseudomonad muconate lactonizing enzyme II, allowing them to promiscuously catalyse the activity of O-succinylbenzoate synthase (menC). Conversely, promiscuity can be decreased as was the case of γ-humulene synthase (a sesquiterpene synthase) from Abies grandis that is known to produce 52 different sesquiterpenes from farnesyl diphosphate upon several mutations.

Studies on enzymes with broad-specificity—not promiscuous, but conceptually close—such as mammalian trypsin and chymotrypsin, and the bifunctional isopropylmalate isomerase/homoaconitase from Pyrococcus horikoshii have revealed that active site loop mobility contributes substantially to the catalytic elasticity of the enzyme.

Toxicity

A promiscuous activity is a non-native activity the enzyme did not evolve to do, but arises due to an accommodating conformation of the active site. However, the main activity of the enzyme is a result not only of selection towards a high catalytic rate towards a particular substrate to produce a particular product, but also to avoid the production of toxic or unnecessary products. For example, if a tRNA synthases loaded an incorrect amino acid onto a tRNA, the resulting peptide would have unexpectedly altered properties, consequently to enhance fidelity several additional domains are present. Similar in reaction to tRNA synthases, the first subunit of tyrocidine synthetase (tyrA) from Bacillus brevis adenylates a molecule of phenylalanine in order to use the adenyl moiety as a handle to produce tyrocidine, a cyclic non-ribosomal peptide. When the specificity of enzyme was probed, it was found that it was highly selective against natural amino acids that were not phenylalanine, but was much more tolerant towards unnatural amino acids. Specifically, most amino acids were not catalysed, whereas the next most catalysed native amino acid was the structurally similar tyrosine, but at a thousandth as much as phenylalanine, whereas several unnatural amino acids where catalysed better than tyrosine, namely D-phenylalanine, β-cyclohexyl-L-alanine, 4-amino-L-phenylalanine and L-norleucine.

One peculiar case of selected secondary activity are polymerases and restriction endonucleases, where incorrect activity is actually a result of a compromise between fidelity and evolvability. For example, for restriction endonucleases incorrect activity (star activity) is often lethal for the organism, but a small amount allows new functions to evolve against new pathogens.

Plant Secondary Metabolism

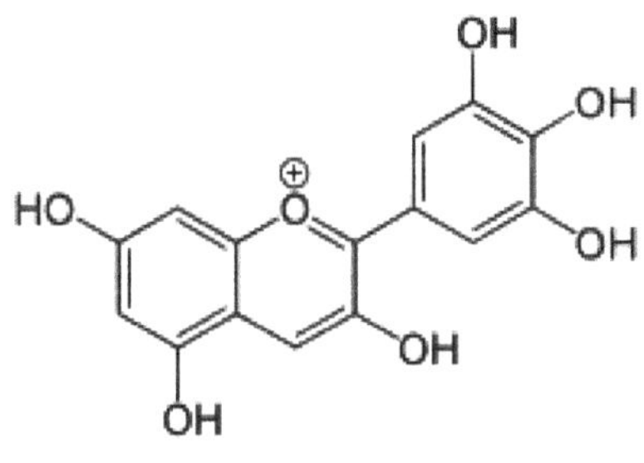

Anthocyanins (delphindin pictured) confer plants, particularly their flowers, with a variety of colours to attract pollinators and a typical example of plant secondary metabolite.

Plants produce a large number of secondary metabolites thanks to enzymes that, unlike those involved in primary metabolism, are less catalytically efficient but have a larger mechanistic elasticity (reaction types) and broader specificities. The liberal drift threshold (caused by the low selective pressure due the small population size) allows the fitness gain endowed by one of the products to maintain the other activities even though they may be physiologically useless.

Biocatalysis

In biocatalysis, many reactions are sought that are absent in nature. To do this, enzymes with a small promiscuous activity towards the required reaction are identified and evolved via directed evolution or rational design.

An example of a commonly evolved enzyme is ω-transaminase which can replace a ketone with a chiral amine and consequently libraries of different homologues are commercially available for rapid biomining (eg. Codexis).

Another example is the possibility of using the promiscuous activities of cysteine synthase (cysM) towards nucleophiles to produce non-proteinogenic amino acids.

Reaction Similarity

Similarity between enzymatic reactions (EC) can be calculated by using bond changes, reaction centres or substructure metrics (EC-BLAST).

Drugs and Promiscuity

Whereas promiscuity is mainly studied in terms of standard enzyme kinetics, drug binding and subsequent reaction is a promiscuous activity as the enzyme catalyses an inactivating reaction towards a novel substrate it did not evolve to catalyse.

Mammalian xenobiotic metabolism, on the other hand, was evolved to have a broad specificity to oxidise, bind and eliminate foreign lipophilic compounds which may be toxic, such as plant alkaloids, so their ability to detoxify anthropogenic xenobiotics is an extension of this.

Diffusion Limited Enzyme

A Diffusion limited enzyme is an enzyme which catalyses a reaction so efficiently that the rate limiting step is that of substratediffusion into the active site, or product diffusion out. This is also known as kinetic perfection or catalytic perfection. Since the rate of catalysis of such enzymes is set by the diffusion-controlled reaction, it therefore represents an intrinsic, physical constraint on evolution (a maximum peak height in the fitness landscape). Diffusion limited perfect enzymes are very rare. Most enzymes catalyse their reactions to a rate that is 1,000-10,000 times slower than this limit. This is due to both the chemical limitations of difficult reactions, and the evolutionary limitations that such high reaction rates do not confer any extra fitness.

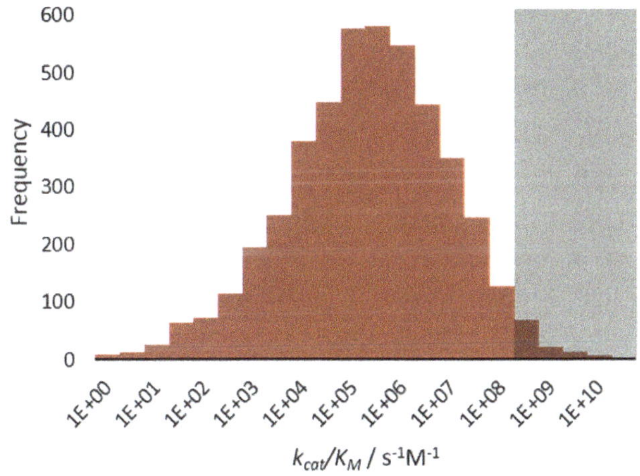

The distribution of known enzyme catalytic rates (k_{cat}/K_M). Most enzymes have a rate around $10^5 s^{-1}M^{-1}$. The fastest enzymes in the dark box on the right ($>10^8\ s^{-1}M^{-1}$) are constrained by the diffusion limit. (Data adapted from reference)

The theory of diffusion-controlled reaction was originally utilized by R.A. Alberty, Gordon Hammes, and Manfred Eigen to estimate the upper limit of enzyme-substrate reaction. According to their estimation, the upper limit of enzyme-substrate reaction was 109 M−1 s−1.

In 1972, it was observed that in the dehydration of H2CO3 catalyzed by carbonic anhydrase, the second-order rate constant obtained experimentally was about 1.5 × 1010 M−1 s−1, which was one order of magnitude higher than the upper limit estimated by Alberty, Hammes, and Eigen based on a simplified model.

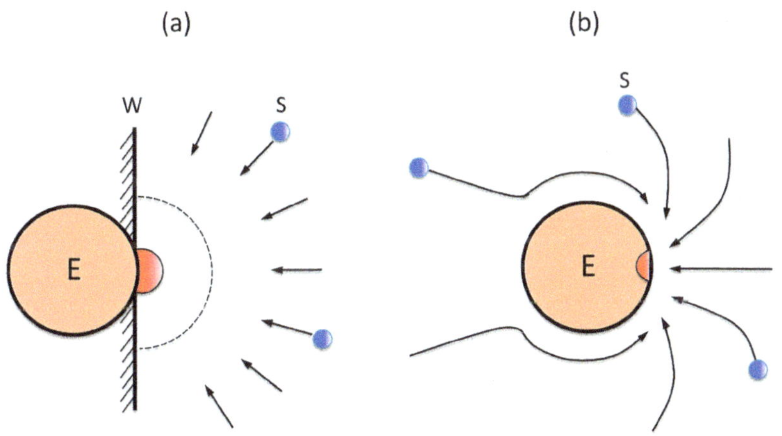

An illustration to show (a) Alberty-Hammes-Eigen model, and (b) Chou's model, where E denotes the enzyme whose active site is colored in red, while the substrate S in blue.

To address such a paradox, Prof. Kuo-Chen Chou and his co-workers proposed a model by taking into account the spatial factor and force field factor between the enzyme and its substrate and found that the upper limit could reach 1010 M−1 s−1, and can be used to explain some surprisingly high reaction rates in molecular biology.

The new upper limit found by Chou et al. for enzyme-substrate reaction was further discussed and analyzed by a series of follow-up studies.

A detailed comparison between the simplified Alberty-Hammes-Eigen's model (a) and the Chou's model (b) in calculating the diffusion-controlled reaction rate of enzyme with its substrate, or the upper limit of enzyme-substrate reaction, was elaborated in the paper.

Mechanism

Kinetically perfect enzymes have a specificity constant, kcat/Km, on the order of 108 to 109 M−1 s−1. The rate of the enzyme-catalysed reaction is limited by diffusion and so the enzyme 'processes' the substrate well before it encounters another molecule.

Some enzymes operate with kinetics which are faster than diffusion rates, which would seem to be impossible. Several mechanisms have been invoked to explain this phenomenon. Some proteins are believed to accelerate catalysis by drawing their substrate in and preorienting them by using dipolar electric fields. Some invoke a quantum-mechanical tunneling explanation whereby a proton or an electron can tunnel through activation barriers, although proton tunneling remains a somewhat controversial idea.

Evolution

It is worth noting that there are not many kinetically perfect enzymes. This can be explained in terms of natural selection. An increase in catalytic speed may be favoured as it could confer some advantage to the organism. However, when the catalytic speed outstrips diffusion speed (i.e. substrates entering and leaving the active site, and also encountering substrates) there is no more advantage to increase the speed even further. The diffusion limit represents an absolute physical constraint on evolution. Increasing the catalytic speed past the diffusion speed will not aid the organism in any way and so represents a global maximum in a fitness landscape. Therefore, these perfect enzymes must have come about by 'lucky' random mutation which happened to spread, or because the faster speed was once useful as part of a different reaction in the enzyme's ancestry.

COFACTOR (BIOCHEMISTRY)

A cofactor is a non-protein chemical compound or metallic ion that is required for a protein's biological activity to happen. These proteins are commonly enzymes, and cofactors can be considered "helper molecules" that assist in biochemical transformations.

Cofactors can be subclassified as either inorganic ions or complex organic molecules called coenzymes, the latter of which is mostly derived from vitamins and other organic essential nutrients in small amounts. A coenzyme that is tightly or even covalently bound is termed a prosthetic group. The two subcategories under coenzyme are cosubstrates and prosthetic groups. Cosubstrates are transiently bound to the protein and will be released at some point, then get back in. The prosthetic groups, on the other hand, are bound permanently to the protein. Both of them have the same function, which is to facilitate the reaction of enzymes and protein. Additionally, some sources also limit the use of the term "cofactor" to inorganic substances. An inactive enzyme without the cofactor is called an apoenzyme, while the complete enzyme with cofactor is called a holoenzyme.

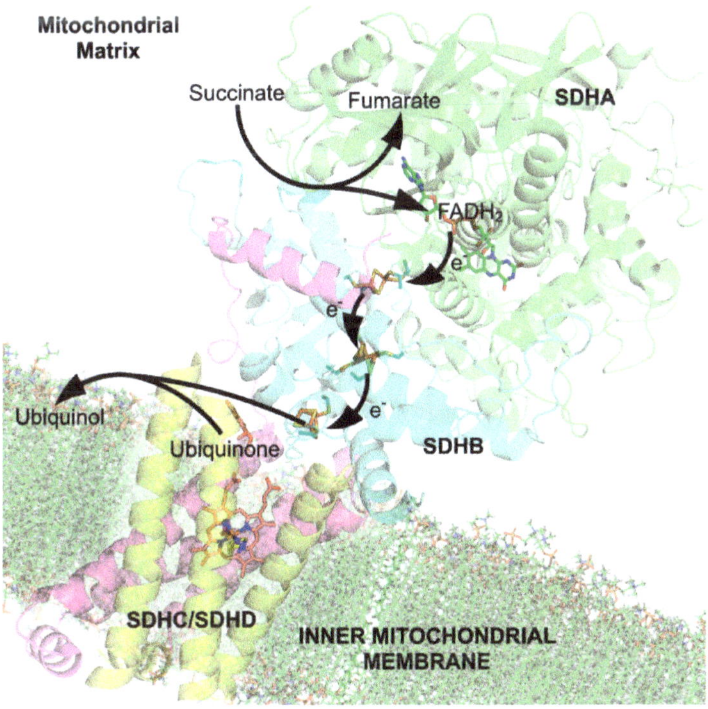

The succinate dehydrogenase complex showing several cofactors, including flavin, iron-sulfur centers, and heme.

Some enzymes or enzyme complexes require several cofactors. For example, the multienzyme complex pyruvate dehydrogenase at the junction of glycolysis and the citric acid cycle requires five organic cofactors and one metal ion: loosely bound thiamine pyrophosphate (TPP), covalently bound lipoamide and flavin adenine dinucleotide (FAD), and the cosubstrates nicotinamide adenine dinucleotide (NAD+) and coenzyme A (CoA), and a metal ion (Mg2+).

Organic cofactors are often vitamins or made from vitamins. Many contain the nucleotide adenosine monophosphate (AMP) as part of their structures, such as ATP, coenzyme A, FAD, and NAD+. This common structure may reflect a common evolutionary origin as part of ribozymes in an ancient RNA world. It has been suggested that the AMP part of the molecule can be considered to be a kind of "handle" by which the enzyme can "grasp" the coenzyme to switch it between different catalytic centers.

Classification

Cofactors can be divided into two broad groups: organic cofactors, such as flavin or heme, and inorganic cofactors, such as the metal ions Mg2+, Cu+, Mn2+, or iron-sulfur clusters.

Organic cofactors are sometimes further divided into coenzymes and prosthetic groups. The term coenzyme refers specifically to enzymes and, as such, to the functional properties of a protein. On the other hand, "prosthetic group" emphasizes the nature of the binding of a cofactor to a protein (tight or covalent) and, thus, refers to a structural property. Different sources give slightly different definitions of coenzymes, cofactors, and prosthetic groups. Some consider tightly bound organic molecules as prosthetic groups and not as coenzymes, while others define all non-protein organic molecules needed for enzyme activity as coenzymes, and classify those that are tightly bound as coenzyme prosthetic groups. It should be noted that these terms are often used loosely.

A 1979 letter in Trends in Biochemical Sciences noted the confusion in the literature and the essentially arbitrary distinction made between prosthetic groups and coenzymes and proposed the following scheme. Here, cofactors were defined as an additional substance apart from protein and substrate that is required for enzyme activity and a prosthetic group as a substance that undergoes its whole catalytic cycle attached to a single enzyme molecule. However, the author could not arrive at a single all-encompassing definition of a "coenzyme" and proposed that this term be dropped from use in the literature.

Inorganic

Metal Ions

Metal ions are common cofactors. The study of these cofactors falls under the area of bioinorganic chemistry. In nutrition, the list of essential trace elements reflects their role as cofactors. In humans this list commonly includes iron, magnesium, manganese, cobalt, copper, zinc, and molybdenum. Although chromium deficiency causes impaired glucose tolerance, no human enzyme that uses this metal as a cofactor has been identified. Iodine is also an essential trace element, but this element is used as part of the structure of thyroid hormones rather than as an enzyme cofactor. Calcium is another special case, in that it is required as a component of the human diet, and it is needed for the full activity of many enzymes, such as nitric oxide synthase, protein phosphatases, and adenylate kinase, but calcium activates these enzymes in allosteric regulation, often binding to these enzymes in a complex with calmodulin. Calcium is, therefore, a cell signaling molecule, and not usually considered a cofactor of the enzymes it regulates.

Other organisms require additional metals as enzyme cofactors, such as vanadium in the nitrogenase of the nitrogen-fixing bacteria of the genus Azotobacter, tungsten in the aldehyde ferredoxin oxidoreductase of the thermophilic archaean Pyrococcus furiosus, and even cadmium in the carbonic anhydrase from the marine diatom Thalassiosira weissflogii.

In many cases, the cofactor includes both an inorganic and organic component. One diverse set of examples is the heme proteins, which consist of a porphyrin ring coordinated to iron.

Ion	Examples of enzymes containing this ion
Cupric	Cytochrome oxidase
Ferrous or Ferric	Catalase Cytochrome (via Heme) Nitrogenase Hydrogenase
Magnesium	Glucose 6-phosphatase Hexokinase DNA polymerase
Manganese	Arginase
Molybdenum	Nitrate reductase Nitrogenase
Nickel	Urease
Zinc	Alcohol dehydrogenase Carbonic anhydrase DNA polymerase

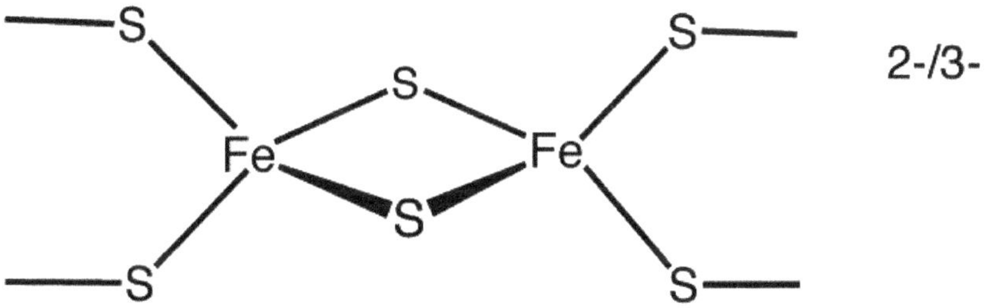

A simple [Fe_2S_2] cluster containing two iron atoms and two sulfur atoms, coordinated by four protein cysteine residues.

Iron-Sulfur Clusters

Iron-sulfur clusters are complexes of iron and sulfur atoms held within proteins by cysteinyl residues. They play both structural and functional roles, including electron transfer, redox sensing, and as structural modules.

Organic

Organic cofactors are small organic molecules (typically a molecular mass less than 1000 Da) that can be either loosely or tightly bound to the enzyme and directly participate in the reaction. In the latter case, when it is difficult to remove without denaturing the enzyme, it can be called a prosthetic group. It is important to emphasize that there is no sharp division between loosely and tightly bound cofactors. Indeed, many such as NAD+ can be tightly bound in some enzymes, while it is loosely bound in others. Another example is thiamine pyrophosphate (TPP), which is tightly bound in transketolase or pyruvate decarboxylase, while it is less tightly bound in pyruvate dehydrogenase. Other coenzymes, flavin adenine dinucleotide (FAD), biotin, and lipoamide, for instance, are covalently bound. Tightly bound cofactors are, in general, regenerated during the same reaction cycle, while loosely bound cofactors can be regenerated in a subsequent reaction catalyzed by a different enzyme. In the latter case, the cofactor can also be considered a substrate or cosubstrate.

Vitamins can serve as precursors to many organic cofactors (e.g., vitamins B1, B2, B6, B12, niacin, folic acid) or as coenzymes themselves (e.g., vitamin C). However, vitamins do have other functions in the body. Many organic cofactors also contain a nucleotide, such as the electron carriers NAD and FAD, and coenzyme A, which carries acyl groups. Most of these cofactors are found in a huge variety of species, and some are universal to all forms of life. An exception to this wide distribution is a group of unique cofactors that evolved in methanogens, which are restricted to this group of archaea.

Vitamins and Derivatives

Cofactor	Vitamin	Additional component	Chemical group(s) transferred	Distribution
Thiamine pyrophosphate	Thiamine (B1)	None	2-carbon groups, α cleavage	Bacteria, archaea and eukaryotes
NAD+ and NADP+	Niacin (B3)	ADP	Electrons	Bacteria, archaea and eukaryotes
Pyridoxal phosphate	Pyridoxine (B6)	None	Amino and carboxyl groups	Bacteria, archaea and eukaryotes
Methylcobalamin	Vitamin B12	Methyl group	acyl groups	Bacteria, archaea and eukaryotes
Cobalamine	Cobalamine (B12)	None	hydrogen, alkyl groups	Bacteria, archaea and eukaryotes
Biotin	Biotin (H)	None	CO2	Bacteria, archaea and eukaryotes
Coenzyme A	Pantothenic acid (B5)	ADP	Acetyl group and other acyl groups	Bacteria, archaea and eukaryotes
Tetrahydrofolic acid	Folic acid (B9)	Glutamate residues	Methyl, formyl, methylene and formimino groups	Bacteria, archaea and eukaryotes
Menaquinone	Vitamin K	None	Carbonyl group and electrons	Bacteria, archaea and eukaryotes
Ascorbic acid	Vitamin C	None	Electrons	Bacteria, archaea and eukaryotes

Flavin mononucleotide	Riboflavin (B2)	None	Electrons	Bacteria, archaea and eukaryotes
Flavin adenine dinucleotide	Riboflavin (B2)	ADP	Electrons	Bacteria, archaea and eukaryotes
Coenzyme F420	Riboflavin (B2)	Amino acids	Electrons	Methanogens and some bacteria

Non-vitamins

Cofactor	Chemical group(s) transferred	Distribution
Adenosine triphosphate	Phosphate group	Bacteria, archaea and eukaryotes
S-Adenosyl methionine	Methyl group	Bacteria, archaea and eukaryotes
Coenzyme B	Electrons	Methanogens
Coenzyme M	Methyl group	Methanogens
Coenzyme Q	Electrons	Bacteria, archaea and eukaryotes
Cytidine triphosphate	Diacylglycerols and lipid head groups	Bacteria, archaea and eukaryotes
Glutathione	Electrons	Some bacteria and most eukaryotes
Heme	Electrons	Bacteria, archaea and eukaryotes
Lipoamide	Electrons, acyl groups	Bacteria, archaea and eukaryotes
Methanofuran	Formyl group	Methanogens
Molybdopterin	Oxygen atoms	Bacteria, archaea and eukaryotes
Nucleotide sugars	Monosaccharides	Bacteria, archaea and eukaryotes
3'-Phosphoadenosine-5'-phosphosulfate	Sulfate group	Bacteria, archaea and eukaryotes
Pyrroloquinoline quinone	Electrons	Bacteria
Tetrahydrobiopterin	Oxygen atom and electrons	Bacteria, archaea and eukaryotes
Tetrahydromethanopterin	Methyl group	Methanogens

Cofactors as Metabolic Intermediates

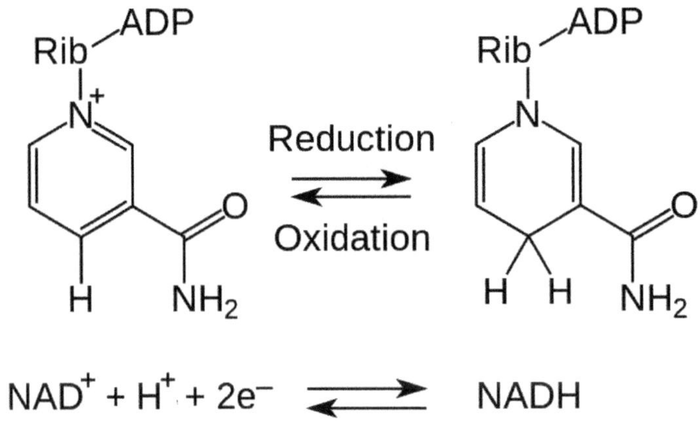

The redox reactions of nicotinamide adenine dinucleotide.

Metabolism involves a vast array of chemical reactions, but most fall under a few basic types of reactions that involve the transfer of functional groups. This common chemistry allows cells to use a small set of metabolic intermediates to carry chemical groups between different reactions. These group-transfer intermediates are the loosely bound organic cofactors, often called coenzymes.

Each class of group-transfer reaction is carried out by a particular cofactor, which is the substrate for a set of enzymes that produce it, and a set of enzymes that consume it. An example of this are the dehydrogenases that use nicotinamide adenine dinucleotide (NAD+) as a cofactor. Here, hundreds of separate types of enzymes remove electrons from their substrates and reduce NAD+ to NADH. This reduced cofactor is then a substrate for any of the reductases in the cell that require electrons to reduce their substrates.

Therefore, these cofactors are continuously recycled as part of metabolism. As an example, the total quantity of ATP in the human body is about 0.1 mole. This ATP is constantly being broken down into ADP, and then converted back into ATP. Thus, at any given time, the total amount of ATP + ADP remains fairly constant. The energy used by human cells requires the hydrolysis of 100 to 150 moles of ATP daily, which is around 50 to 75 kg. In typical situations, humans use up their body weight of ATP over the course of the day. This means that each ATP molecule is recycled 1000 to 1500 times daily.

Evolution

Organic cofactors, such as ATP and NADH, are present in all known forms of life and form a core part of metabolism. Such universal conservation indicates that these molecules evolved very early in the development of living things. At least some of the current set of cofactors may, therefore, have been present in the last universal ancestor, which lived about 4 billion years ago.

Organic cofactors may have been present even earlier in the history of life on Earth. The nucleotide adenosine is present in cofactors that catalyse many basic metabolic reactions such as methyl, acyl, and phosphoryl group transfer, as well as redox reactions. This ubiquitous chemical scaffold has, therefore, been proposed to be a remnant of the RNA world, with early ribozymes evolving to bind a restricted set of nucleotides and related compounds. Adenosine-based cofactors are thought to have acted as interchangeable adaptors that allowed enzymes and ribozymes to bind new cofactors through small modifications in existing adenosine-binding domains, which had originally evolved to bind a different cofactor. This process of adapting a pre-evolved structure for a novel use is referred to as exaptation.

A computational method, IPRO, recently predicted mutations that experimentally switched the cofactor specificity of Candida boidinii xylose reductase from NADPH to NADH.

The first organic cofactor to be discovered was NAD+, which was identified by Arthur Harden and William Youndin 1906. They noticed that adding boiled and filtered yeast extract greatly accelerated alcoholic fermentation in unboiled yeast extracts. They called the unidentified factor responsible for this effect a coferment. Through a long and difficult purification from yeast extracts, this heat-stable factor was identified as a nucleotide sugar phosphate by Hans von Euler-Chelpin. Other cofactors were identified throughout the early 20th century, with ATP being isolated in 1929 by Karl Lohmann, and coenzyme A being discovered in 1945 by Fritz Albert Lipmann.

The functions of these molecules were at first mysterious, but, in 1936, Otto Heinrich Warburg identified the function of NAD+ in hydride transfer. This discovery was followed in the early 1940s by the work of Herman Kalckar, who established the link between the oxidation of sugars and the generation of ATP. This confirmed the central role of ATP in energy transfer that had been proposed by Fritz Albert Lipmann in 1941. Later, in 1949, Morris Friedkin and Albert L. Lehninger proved that NAD+ linked metabolic pathways such as the citric acid cycle and the synthesis of ATP.

Protein-Derived Cofactors

In a number of enzymes, the moiety that acts as a cofactor is formed by post-translational modification of a part of the protein sequence. Examples include tryptophan tryptophylquinone (TTQ), derived from two tryptophan side chains, and 4-methylidene-imidazole-5-one (MIO), derived from an Ala-Ser-Gly motif.

Non-Enzymatic Cofactors

The term is used in other areas of biology to refer more broadly to non-protein (or even protein) molecules that either activate, inhibit, or are required for the protein to function. For example, ligands such as hormones that bind to and activate receptor proteins are termed cofactors or coactivators, whereas molecules that inhibit receptor proteins are termed corepressors. In order to avoid confusion, it has been suggested that such proteins be referred to as coregulators.

ENZYME CATALYSIS

Enzyme catalysis is the increase in the rate of a chemical reaction by the active site of a protein. The protein catalyst (enzyme) may be part of a multi-subunit complex, and/or may transiently or permanently associate with a Cofactor (e.g. adenosine triphosphate). Catalysis of biochemical reactions in the cell is vital due to the very low reaction rates of the uncatalysed reactions at room temperature and pressure. A key driver of protein evolution is the optimization of such catalytic activities via protein dynamics.

The mechanism of enzyme catalysis is similar in principle to other types of chemical catalysis. By providing an alternative reaction route the enzyme reduces the energy required to reach the highest energy transition state of the reaction. The reduction of activation energy (Ea) increases the amount of reactant molecules that achieve a sufficient level of energy, such that they reach the activation energy and form the product. As with other catalysts, the enzyme is not consumed during the reaction (as a substrate is) but is recycled such that a single enzyme performs many rounds of catalysis.

Induced Fit

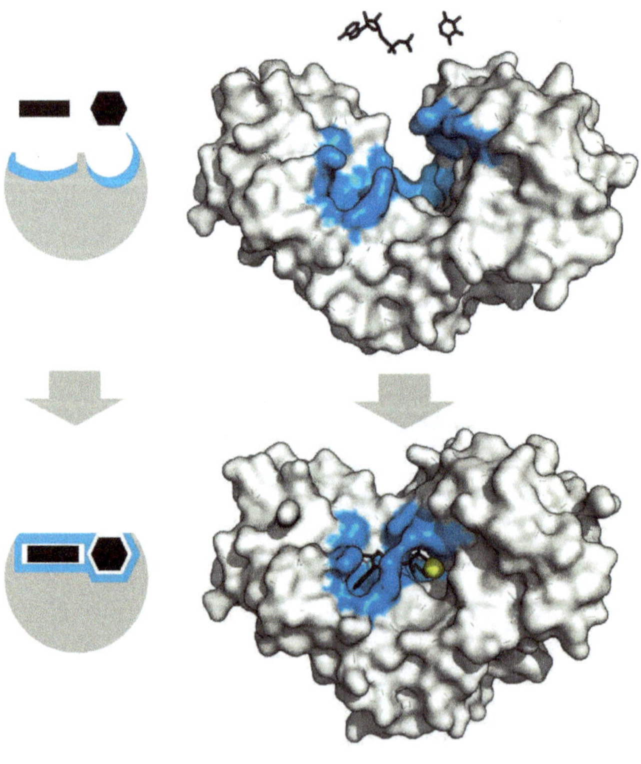

Enzyme changes shape by induced fit upon substrate binding to form enzyme-substrate complex. Hexokinase has a large induced fit motion that closes over the substrates adenosine triphosphate and xylose. Binding sites in blue, substrates in black and Mg^{2+} cofactor in yellow. (PDB: 2E2N, 2E2Q)

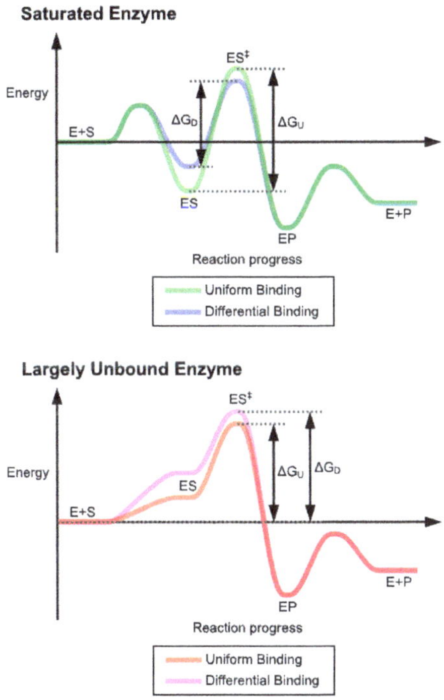

The different mechanisms of substrate binding.

The favored model for the enzyme-substrate interaction is the induced fit model. This model proposes that the initial interaction between enzyme and substrate is relatively weak, but that these weak interactions rapidly induce conformational changes in the enzyme that strengthen binding.

The advantages of the induced fit mechanism arise due to the stabilizing effect of strong enzyme binding. There are two different mechanisms of substrate binding: uniform binding, which has strong substrate binding, and differential binding, which has strong transition state binding. The stabilizing effect of uniform binding increases both substrate and transition state binding affinity, while differential binding increases only transition state binding affinity. Both are used by enzymes and have been evolutionarily chosen to minimize the activation energy of the reaction. Enzymes that are saturated, that is, have a high affinity substrate binding, require differential binding to reduce the energy of activation, whereas small substrate unbound enzymes may use either differential or uniform binding.

These effects have led to most proteins using the differential binding mechanism to reduce the energy of activation, so most substrates have high affinity for the enzyme while in the transition state. Differential binding is carried out by the induced fit mechanism - the substrate first binds weakly, then the enzyme changes conformation increasing the affinity to the transition state and stabilizing it, so reducing the activation energy to reach it.

It is important to clarify, however, that the induced fit concept cannot be used to rationalize catalysis. That is, the chemical catalysis is defined as the reduction of Ea‡ (when the system is already in the ES‡) relative to Ea‡ in the uncatalyzed reaction in water (without the enzyme). The induced fit only suggests that the barrier is lower in the closed form of the enzyme but does not tell us what the reason for the barrier reduction is.

Induced fit may be beneficial to the fidelity of molecular recognition in the presence of competition and noise via the conformational proofreading mechanism . →→→editor

Mechanisms of an Alternative Reaction Route

These conformational changes also bring catalytic residues in the active site close to the chemical bonds in the substrate that will be altered in the reaction. After binding takes place, one or more mechanisms of catalysis lowers the energy of the reaction's transition state, by providing an alternative chemical pathway for the reaction. There are six possible mechanisms of "over the barrier" catalysis as well as a "through the barrier" mechanism:

Proximity and Orientation

The increases the rate of the reaction as enzyme-substrate interactions align reactive chemical groups and hold them close together. This reduces the entropy of the reactants and thus makes reactions such as ligations or addition reactions more favorable, there is a reduction in the overall loss of entropy when two reactants become a single product.

This effect is analogous to an effective increase in concentration of the reagents. The binding of the reagents to the enzyme gives the reaction intramolecular character, which gives a massive rate increase.

However, the situation might be more complex, since modern computational studies have established that traditional examples of proximity effects cannot be related directly to enzyme entropic effects. Also, the original entropic proposal has been found to largely overestimate the contribution of orientation entropy to catalysis.

Proton Donors or Acceptors

Proton donors and acceptors, i.e. acids and base may donate and accept protons in order to stabilize developing charges in the transition state.This typically has the effect of activating nucleophile and electrophile groups, or stabilizing leaving groups. Histidine is often the residue involved in these acid/base reactions, since it has a pKa close to neutral pH and can therefore both accept and donate protons.

Many reaction mechanisms involving acid/base catalysis assume a substantially altered pKa. This alteration of pKa is possible through the local environment of the residue.

Conditions	Acids	Bases
Hydrophobic environment	Increase pKa	Decrease pKa
Adjacent residues of like charge	Increase pKa	Decrease pKa

Salt bridge (and hydrogen bond) formation	Decrease pKa	Increase pKa

pKa can also be influenced significantly by the surrounding environment, to the extent that residues which are basic in solution may act as proton donors, and vice versa.

It is important to clarify that the modification of the pKa's is a pure part of the electrostatic mechanism. Furthermore, the catalytic effect of the above example is mainly associated with the reduction of the pKa of the oxyanion and the increase in the pKa of the histidine, while the proton transfer from the serine to the histidine is not catalyzed significantly, since it is not the rate determining barrier.

Electrostatic Catalysis

Stabilization of charged transition states can also be by residues in the active site forming ionic bonds (or partial ionic charge interactions) with the intermediate. These bonds can either come from acidic or basic side chains found on amino acids such as lysine, arginine, aspartic acid or glutamic acid or come from metal cofactors such as zinc. Metal ions are particularly effective and can reduce the pKa of water enough to make it an effective nucleophile.

Systematic computer simulation studies established that electrostatic effects give, by far, the largest contribution to catalysis. In particular, it has been found that enzyme provides an environment which is more polar than water, and that the ionic transition states are stabilized by fixed dipoles. This is very different from transition state stabilization in water, where the water molecules must pay with "reorganization energy". In order to stabilize ionic and charged states. Thus, the catalysis is associated with the fact that the enzyme polar groups are preorganized

The magnitude of the electrostatic field exerted by an enzyme's active site has been shown to be highly correlated with the enzyme's catalytic rate enhancement

Binding of substrate usually excludes water from the active site, thereby lowering the local dielectric constant to that of an organic solvent. This strengthens the electrostatic interactions between the charged/polar substrates and the active sites. In addition, studies have shown that the charge distributions about the active sites are arranged so as to stabilize the transition states of the catalyzed reactions. In several enzymes, these charge distributions apparently serve to guide polar substrates toward their binding sites so that the rates of these enzymatic reactions are greater than their apparent diffusion-controlled limits.

Covalent Catalysis

Covalent catalysis involves the substrate forming a transient covalent bond with residues in the enzyme active site or with a cofactor. This adds an additional covalent intermediate to the reaction, and helps to reduce the energy of later transition states of the reaction. The covalent bond must, at a later stage in the reaction, be broken to regenerate the enzyme. This mechanism is utilised by the catalytic triad of enzymes such as proteases like chymotrypsin and trypsin, where an acyl-enzyme intermediate is formed. An alternative mechanism is schiff base formation using the free amine from a lysine residue, as seen in the enzyme aldolase during glycolysis.

Some enzymes utilize non-amino acid cofactors such as pyridoxal phosphate (PLP) or thiamine pyrophosphate (TPP) to form covalent intermediates with reactant molecules. Such covalent intermediates function to reduce the energy of later transition states, similar to how covalent intermediates formed with active site amino acid residues allow stabilization, but the capabilities of cofactors allow enzymes to carryout reactions that amino acid side residues alone could not. Enzymes utilizing such cofactors include the PLP-dependent enzyme aspartate transaminase and the TPP-dependent enzyme pyruvate dehydrogenase.

Rather than lowering the activation energy for a reaction pathway, covalent catalysis provides an alternative pathway for the reaction (via to the covalent intermediate) and so is distinct from true catalysis. For example, the energetics of the covalent bond to the serine molecule in chymotrypsin should be compared to the well-understood covalent bond to the nucleophile in the uncatalyzed solution reaction. A true proposal of a covalent catalysis (where the barrier is lower than the corresponding barrier in solution) would require, for example, a partial covalent bond to the transition state by an enzyme group (e.g., a very strong hydrogen bond), and such effects do not contribute significantly to catalysis.

Metal ion Catalysis

The presence of a metal ion in the active site participates in catalysis by coordinating charge stabilization and shielding. Because of a metal's positive charge, only negative charges can be stabilized through metal ions. Metal ions can also act to ionize water by acting as a Lewis acid. Metal ions may also be agents of oxidation and reduction.

Bond Strain

This is the principal effect of induced fit binding, where the affinity of the enzyme to the transition state is greater than to the substrate itself. This induces structural rearrangements which strain substrate bonds into a position closer to the conformation of the transition state, so lowering the energy difference between the substrate and transition state and helping catalyze the reaction.

However, the strain effect is, in fact, a ground state destabilization effect, rather than transition state stabilization effect. Furthermore, enzymes are very flexible and they cannot apply large strain effect.

In addition to bond strain in the substrate, bond strain may also be induced within the enzyme itself to activate residues in the active site.

Quantum Tunneling

These traditional "over the barrier" mechanisms have been challenged in some cases by models and observations of "through the barrier" mechanisms (quantum tunneling).

Some enzymes operate with kinetics which are faster than what would be predicted by the classical ΔG‡. In "through the barrier" models, a proton or an electron can tunnel through activation barriers. Quantum tunneling for protons has been observed in tryptamine oxidation by aromatic amine dehydrogenase.

Interestingly, quantum tunneling does not appear to provide a major catalytic advantage, since the tunneling contributions are similar in the catalyzed and the uncatalyzed reactions in solution. However, the tunneling contribution (typically enhancing rate constants by a factor of ~1000 compared to the rate of reaction for the classical 'over the barrier' route) is likely crucial to the viability of biological organisms. This emphasizes the general importance of tunneling reactions in biology.

In 1971-1972 the first quantum-mechanical model of enzyme catalysis was formulated.

Active Enzyme

The binding energy of the enzyme-substrate complex cannot be considered as an external energy which is necessary for the substrate activation. The enzyme of high energy content may firstly transfer some specific energetic group X1 from catalytic site of the enzyme to the final place of the first bound reactant, then another group X2 from the second bound reactant (or from the second group of the single reactant) must be transferred to active site to finish substrate conversion to product and enzyme regeneration.

We can present the whole enzymatic reaction as a two coupling reactions:

$$S_1 + EX_1 \longrightarrow S_1EX_1 \longrightarrow P_1 + EP_2$$

$$S_2 + EP_2 \longrightarrow S_2EP_2 \longrightarrow P_2 + EX_2$$

It may be seen from reaction (1) that the group X1 of the active enzyme appears in the product due to possibility of the exchange reaction inside enzyme to avoid both electrostatic inhibition and repulsion of atoms. So we represent the active enzyme as a powerful reactant of the enzymatic reaction. The reaction (2) shows incomplete conversion of the substrate because its group X2 remains inside enzyme. This approach as idea had formerly proposed relying on the hypothetical extremely high enzymatic conversions (catalytically perfect enzyme).

The crucial point for the verification of the present approach is that the catalyst must be a complex of the enzyme with the transfer group of the reaction. This chemical aspect is supported by the well-studied mechanisms of the several enzymatic reactions. Let us consider the reaction of peptide bond hydrolysis catalyzed by a pure protein α-chymotrypsin (an enzyme acting without a cofactor), which is a well-studied member of the serine proteases family, see.

We present the experimental results for this reaction as two chemical steps:

$$S_1 + EH \longrightarrow P_1 + EP_2$$

$$EP_2 + H{-}O{-}H \longrightarrow EH + P_2$$

where S1 is a polypeptide, P1 and P2 are products. The first chemical step (3) includes the formation of a covalent acyl-enzyme intermediate. The second step (4) is the deacylation step. It is important to note that the group H+, initially found on the enzyme, but not in water, appears in the product before the step of hydrolysis, therefore it may be considered as an additional group of the enzymatic reaction.

Thus, the reaction (3) shows that the enzyme acts as a powerful reactant of the reaction. According to the proposed concept, the H transport from the enzyme promotes the first reactant conversion, breakdown of the first initial chemical bond (between groups P1 and P2). The step of hydrolysis leads to a breakdown of the second chemical bond and regeneration of the enzyme.

The proposed chemical mechanism does not depend on the concentration of the substrates or products in the medium. However, a shift in their concentration mainly causes free energy changes in the first and final steps of the reactions (1) and (2) due to the changes in the free energy content of every molecule, whether S or P, in water solution. This approach is in accordance with the following mechanism of muscle contraction. The final step of ATP hydrolysis in skeletal muscle is the product release caused by the association of myosin heads with actin. The closing of the actin-binding cleft during the association reaction is structurally coupled with the opening of the nucleotide-binding pocket on the myosin active site.

Notably, the final steps of ATP hydrolysis include the fast release of phosphate and the slow release of ADP. The release of a phosphate anion from bound ADP anion into water solution may be considered as an exergonic reaction because the phosphate anion has low molecular mass.

Thus, we arrive at the conclusion that the primary release of the inorganic phosphate H2PO4− leads to transformation of a significant part of the free energy of ATP hydrolysis into the kinetic energy of the solvated phosphate, producing active streaming. This assumption of a local mechano-chemical transduction is in accord with Tirosh's mechanism of muscle contraction, where the muscle force derives from an integrated action of active streaming created by ATP hydrolysis.

Examples of Catalytic Mechanisms

In reality, most enzyme mechanisms involve a combination of several different types of catalysis.

Triose Phosphate Isomerase

Triose phosphate isomerase (EC 5.3.1.1) catalyses the reversible interconvertion of the two triose phosphates isomers dihydroxyacetone phosphate and D-glyceraldehyde 3-phosphate.

Trypsin

Trypsin (EC 3.4.21.4) is a serine protease that cleaves protein substrates after lysine or arginine residues using a catalytic triad to perform covalent catalysis, and an oxyanion hole to stabilise charge-buildup on the transition states.

Aldolase

Aldolase (EC 4.1.2.13) catalyses the breakdown of fructose 1,6-bisphosphate (F-1,6-BP) into glyceraldehyde 3-phosphate and dihydroxyacetone phosphate (DHAP).

Enzyme Diffusivity

The advent of single-molecule studies led in the 2010s to the observation that the movement of untethered enzymes increases with increasing substrate concentration and increasing reaction enthalpy. Subsequent observations suggest that this increase in diffusivity is driven by transient displacement of the enzyme's center of mass, resulting in a "recoil effect that propels the enzyme".

Reaction Similarity

Similarity between enzymatic reactions (EC) can be calculated by using bond changes, reaction centres or substructure metrics (EC-BLAST).

REFERENCES

1. Nelson, David (2008). <Lehninger Principles of Biochemistry>. New York: W.H. Freeman and Company. p. 184.
2. "coenzymes and cofactors". Retrieved 2007-11-17.
3. "Enzyme Cofactors". Retrieved 2007-11-17.
4. Sauke, David J.; Metzler, David E.; Metzler, Carol M. (2001). Biochemistry: the chemical reactions of living cells (2nd ed.). San Diego: Harcourt/Academic Press. ISBN 0-12-492540-5.
5. Jordan, Frank; Patel, Mulchand S. (2004). Thiamine: catalytic mechanisms in normal and disease states. New York, N.Y: Marcel Dekker. p. 588. ISBN 0-8247-4062-9.
6. Denessiouk KA, Rantanen VV, Johnson MS (2001). "Adenine recognition: a motif present in ATP-, CoA-, NAD-, NADP-, and FAD-dependent proteins". Proteins. 44 (3): 282–91. doi:10.1002/prot.1093. PMID 11455601.
7. Bryce CFA (March 1979). "SAM – semantics and misunderstandings". Trends Biochem. Sci. 4 (3): N62. doi:10.1016/0968-0004(79)90255-X.
8. Metal Ions in Life Sciences
9. Aggett PJ (1985). "Physiology and metabolism of essential trace elements: an outline". Clin Endocrinol Metab. 14 (3): 513–43. doi:10.1016/S0300-595X(85)80005-0. PMID 3905079.
10. Stearns DM (2000). "Is chromium a trace essential metal?". BioFactors. 11 (3): 149–62. doi:10.1002/biof.5520110301. PMID 10875302.

11. Vincent JB (1 April 2000). "The biochemistry of chromium". J. Nutr. 130 (4): 715–8. PMID 10736319.
12. Cavalieri RR (1997). "Iodine metabolism and thyroid physiology: current concepts". Thyroid. 7 (2): 177–81. doi:10.1089/thy.1997.7.177. PMID 9133680.
13. Clapham DE (2007). "Calcium signaling". Cell. 131 (6): 1047–58. doi:10.1016/j.cell.2007.11.028. PMID 18083096.
14. Niki I, Yokokura H, Sudo T, Kato M, Hidaka H (1996). "Ca2+ signaling and intracellular Ca2+ binding proteins". J. Biochem. 120 (4): 685–98. doi:10.1093/oxfordjournals.jbchem.a021466. PMID 8947828.
15. Eady RR (1988). "The vanadium-containing nitrogenase of Azotobacter". BioFactors. 1 (2): 111–6. PMID 3076437.
16. Chan MK, Mukund S, Kletzin A, Adams MW, Rees DC (1995). "Structure of a hyperthermophilic tungstopterin enzyme, aldehyde ferredoxin oxidoreductase". Science. 267 (5203): 1463–9. doi:10.1126/science.7878465. PMID 7878465.
17. Lane TW, Morel FM (2000). "A biological function for cadmium in marine diatoms". Proc. Natl. Acad. Sci. U.S.A. 97 (9): 4627–31. doi:10.1073/pnas.090091397. PMC 18283. PMID 10781068.
18. Lane TW, Saito MA, George GN, Pickering IJ, Prince RC, Morel FM (2005). "Biochemistry: a cadmium enzyme from a marine diatom". Nature. 435 (7038): 42. doi:10.1038/435042a. PMID 15875011.
19. Meyer J (February 2008). "Iron-sulfur protein folds, iron-sulfur chemistry, and evolution". J. Biol. Inorg. Chem. 13 (2): 157–70. doi:10.1007/s00775-007-0318-7. PMID 17992543.
20. Palmer, Trevor (1981). Understanding enzymes. New York: Horwood. ISBN 0-85312-307-1.
21. Cox, Michael; Lehninger, Albert L; Nelson, David R. (2000). Lehninger principles of biochemistry (3rd ed.). New York: Worth Publishers. ISBN 1-57259-153-6.
22. Farrell, Shawn O.; Campbell, Mary K. (2009). Biochemistry (6th ed.). Pacific Grove: Brooks Cole. ISBN 0-495-39041-0.
23. Bolander FF (2006). "Vitamins: not just for enzymes". Curr Opin Investig Drugs. 7 (10): 912–5. PMID 17086936.
24. Rouvière PE, Wolfe RS (15 June 1988). "Novel biochemistry of methanogenesis". J. Biol. Chem. 263 (17): 7913–6. PMID 3131330.
25. Frank RA, Leeper FJ, Luisi BF (2007). "Structure, mechanism and catalytic duality of thiamine-dependent enzymes". Cell. Mol. Life Sci. 64 (7–8): 892–905. doi:10.1007/s00018-007-6423-5. PMID 17429582.
26. Pollak N, Dölle C, Ziegler M (2007). "The power to reduce: pyridine nucleotides—small molecules with a multitude of functions". Biochem. J. 402 (2): 205–18. doi:10.1042/BJ20061638. PMC 1798440. PMID 17295611.
27. Eliot AC, Kirsch JF (2004). "Pyridoxal phosphate enzymes: mechanistic,

structural, and evolutionary considerations". Annu. Rev. Biochem. 73: 383–415. doi:10.1146/annurev.biochem.73.011303.074021. PMID 15189147.

28. Banerjee R, Ragsdale SW (2003). "The many faces of vitamin B12: catalysis by cobalamin-dependent enzymes". Annu. Rev. Biochem. 72: 209–47. doi:10.1146/annurev.biochem.72.121801.161828. PMID 14527323.

29. Jitrapakdee S, Wallace JC (2003). "The biotin enzyme family: conserved structural motifs and domain rearrangements". Curr. Protein Pept. Sci. 4 (3): 217–29. doi:10.2174/1389203033487199. PMID 12769720.

30. Leonardi R, Zhang YM, Rock CO, Jackowski S (2005). "Coenzyme A: back in action". Prog. Lipid Res. 44 (2–3): 125–53. doi:10.1016/j.plipres.2005.04.001. PMID 15893380.

31. Donnelly JG (2001). "Folic acid". Crit Rev Clin Lab Sci. 38 (3): 183–223. doi:10.1080/20014091084209. PMID 11451208.

32. Søballe B, Poole RK (1999). "Microbial ubiquinones: multiple roles in respiration, gene regulation and oxidative stress management" (PDF). Microbiology (Reading, Engl.). 145 (8): 1817–30. doi:10.1099/13500872-145-8-1817. PMID 10463148.

33. Linster CL, Van Schaftingen E (2007). "Vitamin C. Biosynthesis, recycling and degradation in mammals". FEBS J. 274 (1): 1–22. doi:10.1111/j.1742-4658.2006.05607.x. PMID 17222174.

34. Joosten V, van Berkel WJ (2007). "Flavoenzymes". Curr Opin Chem Biol. 11 (2): 195–202. doi:10.1016/j.cbpa.2007.01.010. PMID 17275397.

35. Mack M, Grill S (2006). "Riboflavin analogs and inhibitors of riboflavin biosynthesis". Appl. Microbiol. Biotechnol. 71 (3): 265–75. doi:10.1007/s00253-006-0421-7. PMID 16607521.

36. Bugg, Tim (1997). An introduction to enzyme and coenzyme chemistry. Oxford: Blackwell Science. p. 95. ISBN 0-86542-793-3.

37. Chiang P, Gordon R, Tal J, Zeng G, Doctor B, Pardhasaradhi K, McCann P (1996). "S-Adenosylmethionine and methylation". FASEB J. 10 (4): 471–80. PMID 8647346.

38. Noll KM, Rinehart KL, Tanner RS, Wolfe RS (1986). "Structure of component B (7-mercaptoheptanoylthreonine phosphate) of the methylcoenzyme M methylreductase system of Methanobacterium thermoautotrophicum". Proc. Natl. Acad. Sci. U.S.A. 83 (12): 4238–42. doi:10.1073/pnas.83.12.4238. PMC 323707. PMID 3086878.

39. Taylor CD, Wolfe RS (10 August 1974). "Structure and methylation of coenzyme M(HSCH2CH2SO3)". J. Biol. Chem. 249 (15): 4879–85. PMID 4367810.

40. Balch WE, Wolfe RS (1979). "Specificity and biological distribution of coenzyme M (2-mercaptoethanesulfonic acid)". J. Bacteriol. 137 (1): 256–63. PMC 218444. PMID 104960.

41. Crane FL (1 December 2001). "Biochemical functions of coenzyme Q10". Journal of the American College of Nutrition. 20 (6): 591–8. doi:10.1080/07315724.2001.10719063. PMID 11771674.

42. Buchanan; Gruissem, Jones (2000). Biochemistry & molecular biology of plants (1st ed.). American society of plant physiology. ISBN 0-943088-39-9.
43. Grill D, Tausz T, De Kok LJ (2001). Significance of glutathione in plant adaptation to the environment. Springer. ISBN 1-4020-0178-9.
44. Meister A, Anderson ME (1983). "Glutathione". Annu. Rev. Biochem. 52: 711–60. doi:10.1146/annurev.bi.52.070183.003431. PMID 6137189.
45. Wijayanti N, Katz N, Immenschuh S (2004). "Biology of heme in health and disease". Curr. Med. Chem. 11 (8): 981–6. doi:10.2174/0929867043455521. PMID 15078160.
46. Vorholt JA, Thauer RK (1997). "The active species of 'CO2' utilized by formylmethanofuran dehydrogenase from methanogenic Archaea". Eur. J. Biochem. 248 (3): 919–24. doi:10.1111/j.1432-1033.1997.00919.x. PMID 9342247.
47. Mendel RR, Hänsch R (2002). "Molybdoenzymes and molybdenum cofactor in plants". J. Exp. Bot. 53 (375): 1689–98. doi:10.1093/jxb/erf038. PMID 12147719.
48. Mendel RR, Bittner F (2006). "Cell biology of molybdenum". Biochim. Biophys. Acta. 1763 (7): 621–35. doi:10.1016/j.bbamcr.2006.03.013. PMID 16784786.
49. Ginsburg V (1978). "Comparative biochemistry of nucleotide-linked sugars". Prog. Clin. Biol. Res. 23: 595–600. PMID 351635.
50. Negishi M, Pedersen LG, Petrotchenko E, et al. (2001). "Structure and function of sulfotransferases". Arch. Biochem. Biophys. 390 (2): 149–57. doi:10.1006/abbi.2001.2368. PMID 11396917.
51. Salisbury SA, Forrest HS, Cruse WB, Kennard O (August 1979). "A novel coenzyme from bacterial primary alcohol dehydrogenases". Nature. 280 (5725): 843–4. doi:10.1038/280843a0. PMID 471057.
52. Thony B, Auerbach G, Blau N (2000). "Tetrahydrobiopterin biosynthesis, regeneration and functions". Biochem J. 347 (1): 1–16. doi:10.1042/0264-6021:3470001. PMC 1220924. PMID 10727395.
53. DiMarco AA, Bobik TA, Wolfe RS (1990). "Unusual coenzymes of methanogenesis". Annu. Rev. Biochem. 59: 355–94. doi:10.1146/annurev.bi.59.070190.002035. PMID 2115763.
54. Mitchell P (1979). "The Ninth Sir Hans Krebs Lecture. Compartmentation and communication in living systems. Ligand conduction: a general catalytic principle in chemical, osmotic and chemiosmotic reaction systems". Eur J Biochem. 95 (1): 1–20. doi:10.1111/j.1432-1033.1979.tb12934.x. PMID 378655.
55. Wimmer M, Rose I (1978). "Mechanisms of enzyme-catalyzed group transfer reactions". Annu Rev Biochem. 47: 1031–78. doi:10.1146/annurev.bi.47.070178.005123. PMID 354490.
56. Di Carlo SE, Collins HL (2001). "Estimating ATP resynthesis during a marathon run: a method to introduce metabolism". Advan. Physiol. Edu. 25 (2): 70–1.
57. Chen X, Li N, Ellington AD (2007). "Ribozyme catalysis of metabolism in the

RNA world". Chem. Biodivers. 4 (4): 633–55. doi:10.1002/cbdv.200790055. PMID 17443876.

58. Koch A (1998). "How did bacteria come to be?". Adv Microb Physiol. 40: 353–99. doi:10.1016/S0065-2911(08)60135-6. PMID 9889982.
59. Ouzounis C, Kyrpides N (1996). "The emergence of major cellular processes in evolution". FEBS Lett. 390 (2): 119–23. doi:10.1016/0014-5793(96)00631-X. PMID 8706840.
60. White HB (1976). "Coenzymes as fossils of an earlier metabolic state". J. Mol. Evol. 7 (2): 101–4. doi:10.1007/BF01732468. PMID 1263263.
61. Saran D, Frank J, Burke DH (2003). "The tyranny of adenosine recognition among RNA aptamers to coenzyme A". BMC Evol. Biol. 3: 26. doi:10.1186/1471-2148-3-26. PMC 317284. PMID 14687414.
62. Jadhav VR, Yarus M (2002). "Coenzymes as coribozymes". Biochimie. 84 (9): 877–88. doi:10.1016/S0300-9084(02)01404-9. PMID 12458080.
63. Khoury, GA; Fazelinia, H; Chin, JW; Pantazes, RJ; Cirino, PC; Maranas, CD (October 2009). "Computational design of Candida boidinii xylose reductase for altered cofactor specificity". Protein Science. 18 (10): 2125–38. doi:10.1002/pro.227. PMC 2786976. PMID 19693930
64. Harden A, Young WJ (24 October 1906). "The Alcoholic Ferment of Yeast-Juice" (PDF). Proceedings of the Royal Society B: Biological Sciences. 78 (526): 369–75. doi:10.1098/rspb.1906.0070.
65. "Fermentation of sugars and fermentative enzymes: Nobel Lecture, May 23, 1930" (PDF). Nobel Foundation. Retrieved 2007-09-30.
66. Lohmann K (August 1929). "Über die Pyrophosphatfraktion im Muskel". Naturwissenschaften. 17 (31): 624–5. doi:10.1007/BF01506215.
67. Lipmann F (1 September 1945). "Acetylation of sulfanilamide by liver homogenates and extracts". J. Biol. Chem. 160 (1): 173–90.
68. Warburg O, Christian W (1936). "Pyridin, the hydrogen-transferring component of the fermentation enzymes (pyridine nucleotide)". Biochemische Zeitschrift. 287: 291. doi:10.1002/hlca.193601901199.
69. Kalckar HM (1974). "Origins of the concept oxidative phosphorylation". Mol. Cell. Biochem. 5 (1–2): 55–63. doi:10.1007/BF01874172. PMID 4279328.
70. Lipmann F (1941). "Metabolic generation and utilization of phosphate bond energy". Adv Enzymol. 1: 99–162. doi:10.4159/harvard.9780674366701.c141.
71. Friedkin M, Lehninger AL (1949). "Esterification of inorganic phosphate coupled to electron transport between dihydrodiphosphopyridine nucleotide and oxygen". J. Biol. Chem. 178 (2): 611–23. PMID 18116985.
72. Davidson VL, Wilmot CM (2013). "Posttranslational biosynthesis of the protein-derived cofactor tryptophan tryptophylquinone.". Annu. Rev. Biochem. 82: 531–50. doi:10.1146/annurev-biochem-051110-133601. PMC 4082410. PMID 23746262.

73. Huang SX, Lohman JR, Huang T, Shen B (May 2013). "A new member of the 4-methylideneimidazole-5-one-containing aminomutase family from the enediyne kedarcidin biosynthetic pathway.". Proc. Natl. Acad. Sci. U.S.A. 110 (20): 8069–74. doi:10.1073/pnas.1304733110. PMC 3657804. PMID 23633564.
74. McKenna NJ, O'Malley BW (October 2008). "Editorial: Coactivators and Corepressors: What's in a Name?.". Mol. Endocrinol. 22 (10): 2213–4. doi:10.1210/me.2008-0201. PMC 2582534 . PMID 18701638.
75. Kamerlin, S. C.; Warshel, A (2010). "At the dawn of the 21st century: Is dynamics the missing link for understanding enzyme catalysis?". Proteins: Structure, Function, and Bioinformatics. 78 (6): 1339–75. doi:10.1002/prot.22654. PMC 2841229. PMID 20099310.
76. Koshland DE (February 1958). "Application of a Theory of Enzyme Specificity to Protein Synthesis". Proc. Natl. Acad. Sci. U.S.A. 44 (2): 98–104. Bibcode:1958PNAS...44...98K. doi:10.1073/pnas.44.2.98. PMC 335371. PMID 16590179.
77. Anslyn, E.V.; Dougherty, D.A. (2006). Modern Physical Organic Chemistry. University Science Books. ISBN 1-891389-31-9.
78. Savir Y; Tlusty T (2007). Scalas, Enrico, ed. "Conformational Proofreading: The Impact of Conformational Changes on the Specificity of Molecular Recognition" (PDF). PLOS ONE. 2 (5): e468. Bibcode:2007PLoSO...2..468S. doi:10.1371/journal.pone.0000468. PMC 1868595. PMID 17520027.
79. Stanton, R.V.; Perakyla, M.; Bakowies, D.; Kollman, P.A. (1998). "Combined ab initio and Free Energy Calculations To Study Reactions in Enzymes and Solution: Amide Hydrolysis in Trypsin and Aqueous Solution". J. Am. Chem. Soc. 120 (14): 3448–3457. doi:10.1021/ja972723x.
80. Kuhn, B.; Kollman, P.A. (2000). "QM-FE and Molecular Dynamics Calculations on Catechol O-Methyltransferase: Free Energy of Activation in the Enzyme and in Aqueous Solution and Regioselectivity of the Enzyme-Catalyzed Reaction". J. Am. Chem. Soc. 122 (11): 2586–2596. doi:10.1021/ja992218v.
81. Bruice, T.C.; Lightstone, F.C. (1999). "Ground State and Transition State Contributions to the Rates of Intramolecular and Enzymatic Reactions". Acc. Chem. Res. 32 (2): 127–136. doi:10.1021/ar960131y.
82. Page, M.I.; Jencks, W.P. (1971). "Entropic Contributions to Rate Accelerations in Enzymic and Intramolecular Reactions and the Chelate Effect". Proc. Natl. Acad. Sci. USA. 1971 (68): 1678–1683. Bibcode:1971PNAS...68.1678P. doi:10.1073/pnas.68.8.1678. PMC 389269. PMID 5288752.
83. Warshel, A.; Parson, W.W. (2001). "Dynamics of Biochemical and Biophysical Reactions: Insight from Computer Simulations". Quart. Rev. Biophys. 34: 563–679. doi:10.1017/s0033583501003730.
84. Warshel, A.; Sharma, P.K.; Kato, M.; Xiang, Y.; Liu, H.; Olsson, M.H.M. (2006). "Electrostatic Basis of Enzyme Catalysis". Chem. Rev. 106: 3210–3235. doi:10.1021/cr0503106. PMID 16895325.
85. Warshel, A.; Naray-Szabo, G.; Sussman, F.; Hwang, J.-K. (1989). "How do

Serine Proteases Really Work?". Biochemistry. 1989 (28): 3629–37. doi:10.1021/bi00435a001. PMID 2665806.

86. Marcus, R. A. (1965). "On the Theory of Electron-Transfer Reactions. VI. Unified Treatment for Homogeneous and Electrode Reactions". J. Chem. Phys. 43: 679–701. Bibcode:1965JChPh..43..679M. doi:10.1063/1.1696792.
87. Warshel, A (1978). "Energetics of Enzyme Catalysis". Proc. Natl. Acad. Sci. USA. 75: 5250–5254. Bibcode:1978PNAS...75.5250W. doi:10.1073/pnas.75.11.5250.
88. "How Enzymes Work"
89. "EXTREME ELECTRIC FIELDS POWER CATALYSIS IN THE ACTIVE SITE OF KETOSTEROID ISOMERASE",
90. Toney, M. D. "Reaction specificity in pyridoxal enzymes." Archives of biochemistry and biophysics (2005) 433: 279-287
91. Micronutrient Information Center, Oregon State University
92. Voet, Donald; Judith Voet (2004). Biochemistry. John Wiley & Sons Inc. pp. 986–989. ISBN 0-471-25090-2.
93. Voet, Donald; Judith Voet (2004). Biochemistry. John Wiley & Sons Inc. pp. 604–606. ISBN 0-471-25090-2.
94. Piccirilli, Joseph A.; Vyle, Joseph S.; Caruthers, Marvin H.; Cech, Thomas R. (1993-01-07). "Metal ion catalysis in the Tetrahymena ribozyme reaction". Nature. 361 (6407): 85–88. doi:10.1038/361085a0.
95. Fife, Thomas H.; Przystas, Theodore J. (1985-02-01). "Divalent metal ion catalysis in the hydrolysis of esters of picolinic acid. Metal ion promoted hydroxide ion and water catalyzed reactions". Journal of the American Chemical Society. 107 (4): 1041–1047. doi:10.1021/ja00290a048. ISSN 0002-7863.
96. Stadtman, E. R. (1990-01-01). "Metal ion-catalyzed oxidation of proteins: biochemical mechanism and biological consequences". Free Radical Biology & Medicine. 9 (4): 315–325. doi:10.1016/0891-5849(90)90006-5. ISSN 0891-5849. PMID 2283087.
97. Jencks, William P. (1987) [1969]. Catalysis in Chemistry and Enzymology. McGraw-Hill series in advanced chemistry (reprint ed.). New York: Dover Publications. ISBN 9780486654607.
98. Warshel, A.; Levitt, M. (1976). "Theoretical Studies of Enzymatic Reactions: Dielectric Electrostatic and Steric Stabilization of the Carbonium Ion in the Reaction of Lysozyme". Journal of Molecular Biology. 103 (2): 227–49. doi:10.1016/0022-2836(76)90311-9. PMID 985660.
99. <fundamentals of biochemistry Voet, voet and Pratt 4th edition>, which is similar in shape to the transition state.
100. Garcia-Viloca, M; Gao, J; Karplus, M; Truhlar, DG (2004). "How enzymes work: analysis by modern rate theory and computer simulations". Science. 303 (5655): 186–95. Bibcode:2004Sci...303..186G. doi:10.1126/science.1088172. PMID 14716003.

101. Olsson, MH; Siegbahn, PE; Warshel, A (2004). "Simulations of the large kinetic isotope effect and the temperature dependence of the hydrogen atom transfer in lipoxygenase". Journal of the American Chemical Society. 126 (9): 2820–8. doi:10.1021/ja037233l. PMID 14995199.

102. Masgrau, L; Roujeinikova, A; Johannissen, LO; Hothi, P; Basran, J; Ranaghan, KE; Mulholland, AJ; Sutcliffe, MJ; et al. (2006). "Atomic description of an enzyme reaction dominated by proton tunneling". Science. 312 (5771): 237–41. Bibcode:2006Sci...312..237M. doi:10.1126/science.1126002. PMID 16614214.

103. Hwang, J.-K.; Warshel, A. (1996). "How important are quantum mechanical nuclear motions in enzyme catalysis". J. Am. Chem. Soc. 118: 11745–11751. doi:10.1021/ja962007f.

104. Ball, P. (2004). "Enzymes: By chance, or by design?". Nature. 431 (7007): 396–397. Bibcode:2004Natur.431..396B. doi:10.1038/431396a.

105. Olsson, M.H.M.; Parson, W.W.; Warshel, A. (2006). "Dynamical Contributions to Enzyme Catalysis: Critical Tests of A Popular Hypothesis". Chem. Rev. 106 (5): 1737–1756. doi:10.1021/cr040427e. PMID 16683752.

106. Volkenshtein M.V., Dogonadze R.R., Madumarov A.K., Urushadze Z.D., Kharkats Yu.I. Theory of Enzyme Catalysis.- Molekuliarnaya Biologia, Moscow, 6, 1972, 431-439

107. Volkenshtein M.V., Dogonadze R.R., Madumarov A.K., Urushadze Z.D., Kharkats Yu.I. Electronic and Conformational Interactions in Enzyme Catalysis. In: E.L. Andronikashvili (Ed.), Konformatsionnie Izmenenia Biopolimerov v Rastvorakh, Publishing House "Nauka", Moscow, 1973, 153-157

108. Foigel, A.G. (2011). "Is the enzyme a powerful reactant of the biochemical reaction?". Mol. Cell. Biochem 352: 87-89 Foigel, Alexander G. (2011). "Is the enzyme a powerful reactant of the biochemical reaction?". Molecular and Cellular Biochemistry. 352 (1–2): 87–9. doi:10.1007/s11010-011-0742-4. PMID 21318350.

109. Fogel, A.G. (1982). "Cooperativity of enzymatic reactions and molecular aspects of energy transduction". Mol. Cell. Biochem. 47: 59–64. doi:10.1007/bf00241567.

110. Hengge, AC; Stein, RL (2004). "Role of protein conformational mobility in enzyme catalysis: acylation of alpha-chymotrypsin by specific peptide substrates". Biochemistry. 43: 742–747. doi:10.1021/bi030222k.

111. Lymn, RW; Taylor, EW. (1971). "Mechanism of adenosine triphosphate hydrolysis by actomyosin". Biochemistry. 10: 4617–4624. doi:10.1021/bi00801a004.

112. Holmes, KC; Angert, I; Kull, FG; Jahn, W; Schroder, RR. (2003). "Electron cryo-microscopy shows how strong binding of myosin to actin releases nucleotide". Nature. 425: 423–427. Bibcode:2003Natur.425..423H. doi:10.1038/nature02005.

113. Siemankowski, RF; Wiseman, MO; White, HD. (1985). "ADP dissociation from actomyosin subfragment 1 is sufficiently slow to limit the unloaded shortening velocity in vertebrate muscle". Proc. Natl. Acad. Sci. USA. 82: 658–662. Bibcode:1985PNAS...82..658S. doi:10.1073/pnas.82.3.658.

114. White, HD; Belknap, B; Webb, MR. (1997). "Kinetics of nucleoside triphosphate cleavage and phosphate release steps by associated rabbit skeletal actomyosin, measured using a novel fluorescent probe for phosphate". Biochemistry. 36: 11828–11836. doi:10.1021/bi970540h.

115. Tirosh, R; Low, WZ; Oplatka, A. (1990). "Translational motion of actin filaments in the presence of heavy meromyosin and MgATP as measured by Doppler broadening of laser light scattering". Biochim. Biophys. Acta. 1037: 274–280. doi:10.1016/0167-4838(90)90025-b.

116. Tirosh, R. (2006). "Ballistic protons and microwave-induced water solutions (solitons) in bioenergetic transformations". Int. J. Mol. Sci. 7: 320–345. doi:10.3390/i7090320.

117. Muddana, Hari S.; Sengupta, Samudra; Mallouk, Thomas E.; et al. (28 January 2010). "Substrate Catalysis Enhances Single-Enzyme Diffusion". Journal of the American Chemical Society. 132 (7): 2110–1. doi:10.1021/ja908773a. PMC 2832858. PMID 20108965.

118. Riedel, Clement; Gabizon, Ronen; Wilson, Christian A. M.; et al. (8 January 2015). "The heat released during catalytic turnover enhances the diffusion of an enzyme". Nature. 517: 227–30. Bibcode:2015Natur.517..227R. doi:10.1038/nature14043. Lay summary – Nature: News & Views (8 January 2015).

119. Rahman, SA; Cuesta, SM; Furnham, N; Holliday, GL; Thornton, JM (2014). "EC-BLAST: a tool to automatically search and compare enzyme reactions". Nature Methods. 11: 171–174. doi:10.1038/nmeth.2803. PMID 24412978.

CHAPTER 5

Enzyme Kinetics

Chapter Contents

- **Introduction**
- **General Principles**
- **Enzyme Assays**
- **Single-Substrate Reactions**
- **Multi-Substrate Reactions**
- **Non-Michaelis–Menten Kinetics**
- **Chemical Mechanism**
- **Enzyme Inhibition And Activation**
- **Mechanisms of Catalysis**
- **History**
- **Software**
- **Hanes–Woolf Plot**
- **Lineweaver–Burk Plot**

Enzymes are protein catalysts that, like all catalysts, speed up the rate of a chemical reaction without being used up in the process.

They achieve their effect by temporarily binding to the substrate and, in doing so, lowering the activation energy needed to convert it to a product.

The rate at which an enzyme works is influenced by several factors, e.g.,

- the concentration of substrate molecules (the more of them available, the quicker the enzyme molecules collide and bind with them). The concentration of substrate is designated [S] and is expressed in units of molarity.
- the temperature. As the temperature rises, molecular motion — and hence collisions between enzyme and substrate — speed up. But as enzymes are proteins, there is an upper limit beyond which the enzyme becomes denatured and ineffective.

- the presence of inhibitors.
    - competitive inhibitors are molecules that bind to the same site as the substrate — preventing the substrate from binding as they do so — but are not changed by the enzyme.
    - noncompetitive inhibitors are molecules that bind to some other site on the enzyme reducing its catalytic power.
- pH. The conformation of a protein is influenced by pH and as enzyme activity is crucially dependent on its conformation, its activity is likewise affected.

INTRODUCTION

Enzyme kinetics is the study of the chemical reactions that are catalysed by enzymes. In enzyme kinetics, the reaction rate is measured and the effects of varying the conditions of the reaction are investigated. Studying an enzyme's kinetics in this way can reveal the catalytic mechanism of this enzyme, its role in metabolism, how its activity is controlled, and how a drug or an agonist might inhibit the enzyme.

Dihydrofolate reductase from E. coli with its two substrates dihydrofolate (right) and NADPH (left), bound in the active site. The protein is shown as a ribbon diagram, with alpha helices in red, beta sheets in yellow and loops in blue. Generated from 7DFR.

Enzymes are usually protein molecules that manipulate other molecules — the enzymes' substrates. These target molecules bind to an enzyme's active site and are transformed into products through a series of steps known as the enzymatic mechanism

E + S ⇄ ES ⇄ ES* ⇄ EP ⇄ E + P

These mechanisms can be divided into single-substrate and multiple-substrate mechanisms. Kinetic studies on enzymes that only bind one substrate, such as triosephosphate isomerase, aim to measure the affinity with which the enzyme binds this substrate and the turnover rate. Some other examples of enzymes are phosphofructokinase and hexokinase, both of which are important for cellular respiration (glycolysis).

When enzymes bind multiple substrates, such as dihydrofolate reductase (shown right), enzyme kinetics can also show the sequence in which these substrates bind and the sequence in which products are released. An example of enzymes that bind a single substrate and release multiple products are proteases, which cleave one protein substrate into two polypeptide products. Others join two substrates together, such as DNA polymerase linking a nucleotide to DNA. Although these mechanisms are often a complex series of steps, there is typically one rate-determining step that determines the overall kinetics. This rate-determining step may be a chemical reaction or a conformational change of the enzyme or substrates, such as those involved in the release of product(s) from the enzyme.

Knowledge of the enzyme's structure is helpful in interpreting kinetic data. For example, the structure can suggest how substrates and products bind during catalysis; what changes occur during the reaction; and even the role of particular amino acid residues in the mechanism. Some enzymes change shape significantly during the mechanism; in such cases, it is helpful to determine the enzyme structure with and without bound substrate analogues that do not undergo the enzymatic reaction.

Not all biological catalysts are protein enzymes; RNA-based catalysts such as ribozymes and ribosomes are essential to many cellular functions, such as RNA splicing and translation. The main difference between ribozymes and enzymes is that RNA catalysts are composed of nucleotides, whereas enzymes are composed of amino acids. Ribozymes also perform a more limited set of reactions, although their reaction mechanisms and kinetics can be analysed and classified by the same methods.

GENERAL PRINCIPLES

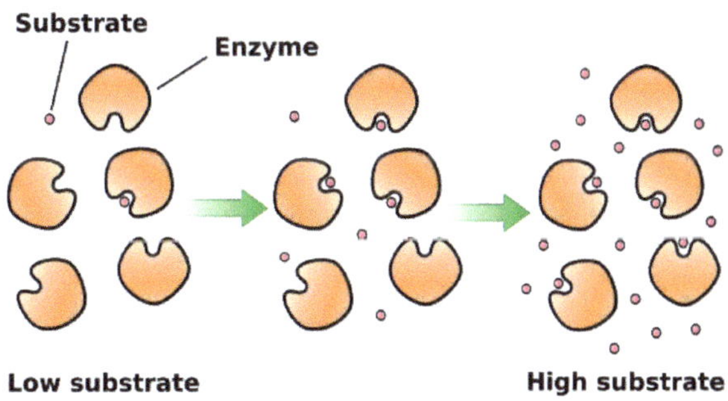

As larger amounts of substrate are added to a reaction, the available enzyme binding sites become filled to the limit of V_{max}. Beyond this limit the enzyme is saturated with substrate and the reaction rate ceases to increase.

The reaction catalysed by an enzyme uses exactly the same reactants and produces exactly the same products as the uncatalysed reaction. Like other catalysts, enzymes do not alter the position of equilibrium between substrates and products.However, unlike uncatalysed chemical reactions, enzyme-catalysed reactions display saturation kinetics. For a given enzyme concentration and for relatively low substrate concentrations, the reaction rate increases linearly with substrate concentration; the enzyme molecules are largely free to catalyse the reaction, and increasing substrate concentration means an increasing rate at which the enzyme and substrate molecules encounter one another. However, at relatively high substrate concentrations, the reaction rate asymptotically approaches the theoretical maximum; the enzyme active sites are almost all occupied and the reaction rate is determined by the intrinsic turnover rate of the enzyme. The substrate concentration midway between these two limiting cases is denoted by KM.

The two most important kinetic properties of an enzyme are how quickly the enzyme becomes saturated with a particular substrate, and the maximum rate it can achieve. Knowing these properties suggests what an enzyme might do in the cell and can show how the enzyme will respond to changes in these conditions.

ENZYME ASSAYS

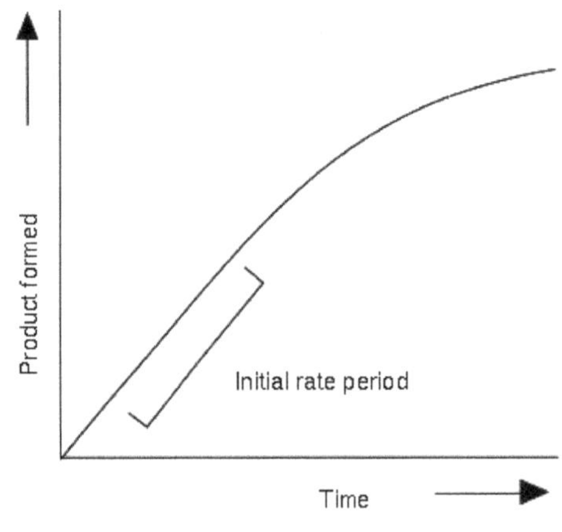

Progress curve for an enzyme reaction. The slope in the initial rate period is the **initial rate of reaction** v. The Michaelis–Menten equation describes how this slope varies with the concentration of substrate.

Enzyme assays are laboratory procedures that measure the rate of enzyme reactions. Because enzymes are not consumed by the reactions they catalyse, enzyme assays usually follow changes in the concentration of either substrates or products to measure the rate of reaction. There are many methods of measurement. Spectrophotometric assays observe change in the absorbance of light between products and reactants; radiometric assays involve the incorporation or release of radioactivity to measure the amount of product made over time. Spectrophotometric assays are most convenient since they allow the rate of the reaction to be measured continuously. Although radiometric assays require the removal and counting of samples (i.e., they are discontinuous assays) they are usually extremely sensitive and can measure very low levels of enzyme activity. An analogous approach is to use mass spectrometry to monitor the incorporation or release of stable isotopes as substrate is converted into product.

The most sensitive enzyme assays use lasers focused through a microscope to observe changes in single enzyme molecules as they catalyse their reactions. These measurements either use changes in the fluorescence of cofactors during an enzyme's reaction mechanism, or of fluorescent dyes added onto specific sites of the protein to report movements that occur during catalysis. These studies are providing a new view of the kinetics and dynamics of single enzymes, as opposed to traditional enzyme kinetics, which observes the average behaviour of populations of millions of enzyme molecules.

An example progress curve for an enzyme assay is shown above. The enzyme produces product at an initial rate that is approximately linear for a short period after the start of the reaction. As the reaction proceeds and substrate is consumed, the rate continuously slows (so long as substrate is not still at saturating levels). To measure the initial (and maximal) rate, enzyme assays are typically carried out while the reaction has progressed only a few percent towards total completion. The length of the initial rate period depends on the assay conditions and can range from milliseconds to hours. However, equipment for rapidly mixing liquids allows fast kinetic measurements on initial rates of less than one second. These very rapid assays are essential for measuring pre-steady-state kinetics, which are discussed below.

Most enzyme kinetics studies concentrate on this initial, approximately linear part of enzyme reactions. However, it is also possible to measure the complete reaction curve and fit this data to a non-linear rate equation. This way of measuring enzyme reactions is called progress-curve analysis. This approach is useful as an alternative to rapid kinetics when the initial rate is too fast to measure accurately.

SINGLE-SUBSTRATE REACTIONS

Enzymes with single-substrate mechanisms include isomerases such as triosephosphateisomerase or bisphosphoglycerate mutase, intramolecular lyases such as adenylate cyclase and the hammerhead ribozyme, an RNA lyase. However, some enzymes that only have a single substrate do not fall into this category of mechanisms. Catalase is an example of this, as the enzyme reacts with a first molecule of hydrogen peroxide substrate, becomes oxidised and is then reduced by a second molecule of substrate. Although a single substrate is involved, the existence of a modified enzyme intermediate means that the mechanism of catalase is actually a ping–pong mechanism, a type of mechanism that is discussed in the Multi-substrate reactions section below.

Michaelis–Menten Kinetics

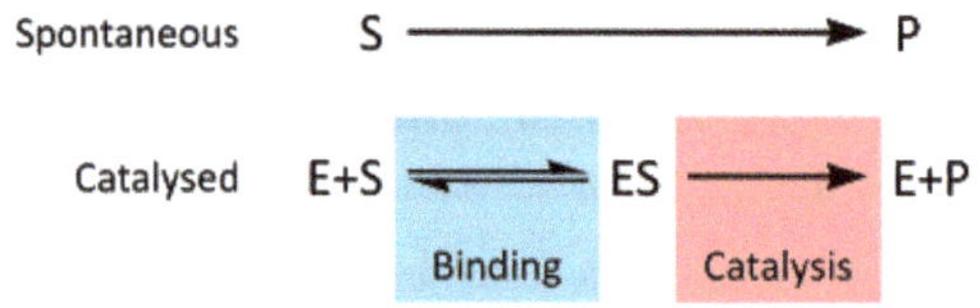

A chemical reaction mechanism with or without enzyme catalysis. The enzyme (E) binds substrate (S) to produce product (P).

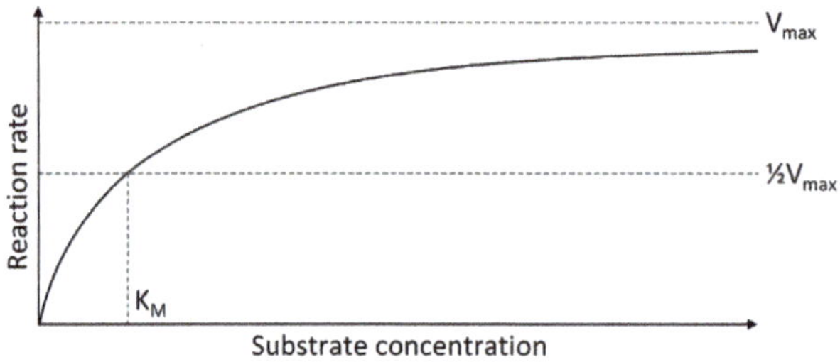

Saturation curve for an enzyme reaction showing the relation between the substrate concentration and reaction rate.

As enzyme-catalysed reactions are saturable, their rate of catalysis does not show a linear response to increasing substrate. If the initial rate of the reaction is measured over a range of substrate concentrations (denoted as [S]), the reaction rate (v) increases as [S] increases, as shown on the right. However, as [S] gets higher, the enzyme becomes saturated with substrate and the rate reaches Vmax, the enzyme's maximum rate.

The Michaelis–Menten kinetic model of a single-substrate reaction is shown on the right. There is an initial bimolecular reaction between the enzyme E and substrate S to form the enzyme–substrate complex ES. The rate of enzymatic reaction increases with the increase of the substrate concentration up to a certain level called Vmax; at Vmax, increase in substrate concentration does not cause any increase in reaction rate as there no more enzyme (E) available for reacting with substrate (S). Here, the rate of reaction becomes dependent on the ES complex and the reaction becomes a unimolecular reaction with an order of zero. Though the enzymatic mechanism for the unimolecular reaction $ES \xrightarrow{k_{cat}} E + P$ can be quite complex, there is typically one rate-determining enzymatic step that allows this reaction to be modelled as a single catalytic step with an apparent unimolecular rate constant kcat. If the reaction path proceeds over one or several intermediates, kcat will be a function of several elementary rate constants, whereas in the simplest case of a single elementary reaction (e.g. no intermediates) it will be identical to the elementary unimolecular rate constant k2. The apparent unimolecular rate constant kcat is also called turnover number and denotes the maximum number of enzymatic reactions catalysed per second.

The Michaelis–Menten equation describes how the (initial) reaction rate v0 depends on the position of the substrate-binding equilibrium and the rate constant k2.

$$v_0 = \frac{V_{\max}[\mathrm{S}]}{K_M + [\mathrm{S}]}$$

with the constants

$$K_M \stackrel{\text{def}}{=} \frac{k_2 + k_{-1}}{k_1} \approx K_D$$

$$V_{\max} \stackrel{\text{def}}{=} k_{cat}[\mathrm{E}]_{tot}$$

This Michaelis–Menten equation is the basis for most single-substrate enzyme kinetics. Two crucial assumptions underlie this equation (apart from the general assumption about the mechanism only involving no intermediate or product inhibition, and there is no allostericity or cooperativity). The first assumption is the so-called quasi-steady-state assumption (or pseudo-steady-state hypothesis), namely that the concentration of the substrate-bound enzyme (and hence also the unbound enzyme) changes much more slowly than those of the product and substrate and thus the change over time of the complex can be set to zero $d[\mathrm{ES}]/dt \stackrel{!}{=} 0$. The second assumption is that the total enzyme concentration does not change over time, thus $[\mathrm{E}]_{\text{tot}} = [\mathrm{E}] + [\mathrm{ES}] \stackrel{!}{=} \text{const}$. A complete derivation can be found here.

The Michaelis constant KM is experimentally defined as the concentration at which the rate of the enzyme reaction is half Vmax, which can be verified by substituting [S] = Km into the Michaelis–Menten equation and can also be seen graphically. If the rate-determining enzymatic step is slow compared to substrate dissociation $(k_2 \ll k_{-1})$, the Michaelis constant KM is roughly the dissociation constant KD of the ES complex.

If $[\mathrm{S}]$ is small compared to K_M then the term $[\mathrm{S}]/(K_M + [\mathrm{S}]) \approx [\mathrm{S}]/K_M$ and also very little ES complex is formed, thus $[\mathrm{E}]_0 \approx [\mathrm{E}]$. Therefore, the rate of product formation is

$$v_0 \approx \frac{k_{cat}}{K_M}[\mathrm{E}][\mathrm{S}] \qquad \text{if } [\mathrm{S}] \ll K_M$$

Thus the product formation rate depends on the enzyme concentration as well as on the substrate concentration, the equation resembles a bimolecular reaction with a corresponding pseudo-second order rate constant k_2/K_M. This constant is a measure of catalytic efficiency. The most efficient enzymes reach a k_2/K_M in the range of 108 – 1010 M−1 s−1. These enzymes are so efficient they effectively catalyse a reaction each time they encounter a substrate molecule and have thus reached an upper theoretical limit for efficiency (diffusion limit); and are sometimes referred to as kinetically perfect enzymes.

Direct use of the Michaelis–Menten Equation for time Course Kinetic Analysis

The observed velocities predicted by the Michaelis–Menten equation can be used to directly model the time course disappearance of substrate and the production of product through incorporation of the Michaelis–Menten equation into the equation for first order chemical kinetics. This can only be achieved however if one recognises the problem associated with the use of Euler's number in the description of first order chemical kinetics. i.e. e−k is a split constant that introduces a systematic error into calculations and can be rewritten as a single constant which represents the remaining substrate after each time period.

$$[S] = [S]_0(1 - k)^t$$

$$[S] = [S]_0(1 - v/[S]_0)^t$$

$$[S] = [S]_0(1 - (V_{max}[S]_0/(K_M + [S]_0)/[S]_0))^t$$

In 1983 Stuart Beal (and also independently Santiago Schnell and Claudio Mendoza in 1997) derived a closed form solution for the time course kinetics analysis of the Michaelis-Menten mechanism. The solution, known as the Schnell-Mendoza equation, has the form:

$$\frac{[S]}{K_M} = W\,[F(t)]$$

where W[] is the Lambert-W function. and where F(t) is

$$F(t) = \frac{[S]_0}{K_M}\exp\left(\frac{[S]_0}{K_M} - \frac{V_{max}}{K_M}t\right)$$

This equation is encompassed by the equation below, obtained by Berberan-Santos (MATCH Commun. Math. Comput. Chem. 63 (2010) 283), which is also valid when the initial substrate concentration is close to that of enzyme,

$$\frac{[S]}{K_M} = W\,[F(t)] - \frac{V_{max}}{k_{cat}K_M}\frac{W\,[F(t)]}{1 + W\,[F(t)]}$$

where W[] is again the Lambert-W function.

Linear Plots of the Michaelis–Menten Equation

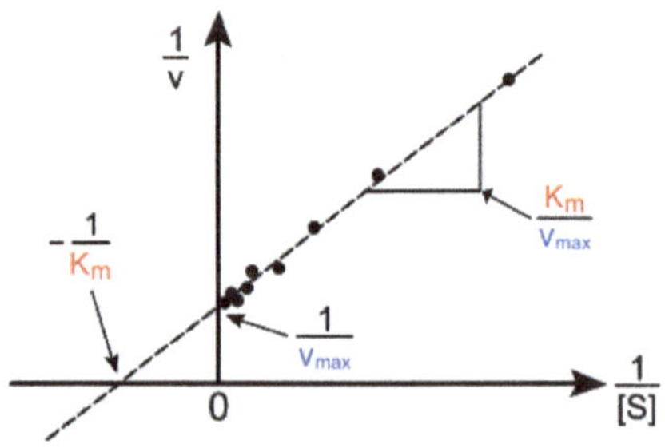

Lineweaver–Burk or double-reciprocal plot of kinetic data, showing the significance of the axis intercepts and gradient.

The plot of v versus [S] above is not linear; although initially linear at low [S], it bends over to saturate at high [S]. Before the modern era of nonlinear curve-fitting on computers, this nonlinearity could make it difficult to estimate KM and Vmaxaccurately. Therefore, several researchers developed linearisations of the Michaelis–Menten equation, such as the Lineweaver–Burk plot, the Eadie–Hofstee diagram and the Hanes–Woolf plot. All of these linear representations can be useful for visualising data, but none should be used to determine kinetic parameters, as computer software is readily available that allows for more accurate determination by nonlinear regression methods.

The Lineweaver–Burk plot or double reciprocal plot is a common way of illustrating kinetic data. This is produced by taking the reciprocal of both sides of the Michaelis–Menten equation. As shown on the right, this is a linear form of the Michaelis–Menten equation and produces a straight line with the equation y = mx + c with a y-intercept equivalent to 1/Vmax and an x-intercept of the graph representing −1/KM.

$$\frac{1}{v} = \frac{K_M}{V_{\max}[S]} + \frac{1}{V_{\max}}$$

Naturally, no experimental values can be taken at negative 1/[S]; the lower limiting value 1/[S] = 0 (the y-intercept) corresponds to an infinite substrate concentration, where 1/v=1/Vmax as shown at the right; thus, the x-intercept is an extrapolation of the experimental data taken at positive concentrations. More generally, the Lineweaver–Burk plot skews the importance of measurements taken at low substrate concentrations and, thus, can yield inaccurate estimates of Vmax and KM. A more accurate linear plotting method is the Eadie-Hofstee plot. In this case, v is plotted against v/[S]. In the third common linear representation, the Hanes-Woolf plot, [S]/v is plotted against [S]. In general, data normalisation can help diminish the amount of experimental work and can increase the reliability of the output, and is suitable for both graphical and numerical analysis.

Practical Significance of Kinetic Constants

The study of enzyme kinetics is important for two basic reasons. Firstly, it helps explain how enzymes work, and secondly, it helps predict how enzymes behave in living organisms. The kinetic constants defined above, KM and Vmax, are critical to attempts to understand how enzymes work together to control metabolism.

Making these predictions is not trivial, even for simple systems. For example, oxaloacetate is formed by malate dehydrogenase within the mitochondrion. Oxaloacetate can then be consumed by citrate synthase, phosphoenolpyruvate carboxykinase or aspartate aminotransferase, feeding into the citric acid cycle, gluconeogenesis or aspartic acid biosynthesis, respectively. Being able to predict how much oxaloacetate goes into which pathway requires knowledge of the concentration of oxaloacetate as well as the concentration and kinetics of each of these enzymes. This aim of predicting the behaviour of metabolic pathways reaches its most complex expression in the synthesis of huge amounts of kinetic and gene expression data into mathematical models of entire organisms. Alternatively, one useful simplification of the metabolic modelling problem is to ignore the underlying enzyme kinetics and only rely on information about the reaction network's stoichiometry, a technique called flux balance analysis.

Michaelis–Menten Kinetics with Intermediate

One could also consider the less simple case

$$E + S \underset{k_{-1}}{\overset{k_1}{\rightleftharpoons}} ES \overset{k_2}{\longrightarrow} EI \overset{k_3}{\longrightarrow} E + P$$

where a complex with the enzyme and an intermediate exists and the intermediate is converted into product in a second step. In this case we have a very similar equation

$$v_0 = k_{cat} \frac{[S][E]_0}{K'_M + [S]}$$

but the constants are different

$$K'_M \stackrel{\text{def}}{=} \frac{k_3}{k_2 + k_3} K_M = \frac{k_3}{k_2 + k_3} \cdot \frac{k_2 + k_{-1}}{k_1}$$

$$k_{cat} \stackrel{\text{def}}{=} \frac{k_3 k_2}{k_2 + k_3}$$

We see that for the limiting case $k_3 \gg k_2$, thus when the last step from $EI \longrightarrow E + P$ is much faster than the previous step, we get again the original equation. Mathematically we have then $K'_M \approx K_M$ and $k_{cat} \approx k_2$.

MULTI-SUBSTRATE REACTIONS

Multi-substrate reactions follow complex rate equations that describe how the substrates bind and in what sequence. The analysis of these reactions is much simpler if the concentration of substrate A is kept constant and substrate B varied. Under these conditions, the enzyme behaves just like a single-substrate enzyme and a plot of v by [S] gives apparent KM and Vmax constants for substrate B. If a set of these measurements is performed at different fixed concentrations of A, these data can be used to work out what the mechanism of the reaction is. For an enzyme that takes two substrates A and B and turns them into two products P and Q, there are two types of mechanism: ternary complex and ping–pong.

Ternary-Complex Mechanisms

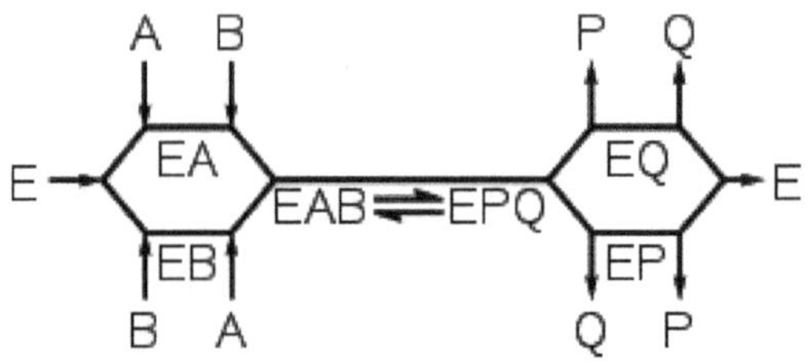

Random-order ternary-complex mechanism for an enzyme reaction. The reaction path is shown as a line and enzyme intermediates containing substrates A and B or products P and Q are written below the line.

In these enzymes, both substrates bind to the enzyme at the same time to produce an EAB ternary complex. The order of binding can either be random (in a random mechanism) or substrates have to bind in a particular sequence (in an ordered mechanism). When a set of v by [S] curves (fixed A, varying B) from an enzyme with a ternary-complex mechanism are plotted in a Lineweaver–Burk plot, the set of lines produced will intersect.

Enzymes with ternary-complex mechanisms include glutathione S-transferase, dihydrofolate reductase and DNA polymerase. The following links show short animations of the ternary-complex mechanisms of the enzymes dihydrofolate reductase[β] and DNA polymerase[γ].

Ping–pong Mechanisms

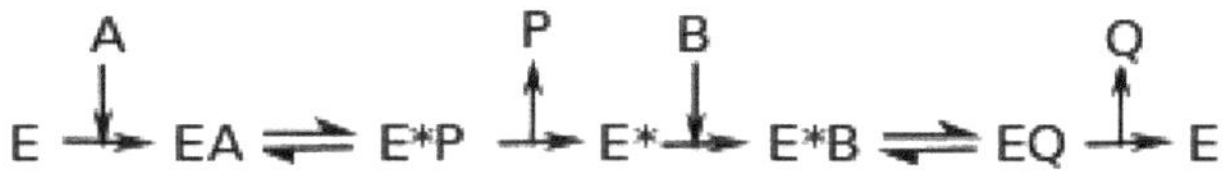

Ping–pong mechanism for an enzyme reaction. Intermediates contain substrates A and B or products P and Q.

As shown on the right, enzymes with a ping-pong mechanism can exist in two states, E and a chemically modified form of the enzyme E*; this modified enzyme is known as an intermediate. In such mechanisms, substrate A binds, changes the enzyme to E* by, for example, transferring a chemical group to the active site, and is then released. Only after the first substrate is released can substrate B bind and react with the modified enzyme, regenerating the unmodified E form. When a set of v by [S] curves (fixed A, varying B) from an enzyme with a ping–pong mechanism are plotted in a Lineweaver–Burk plot, a set of parallel lines will be produced. This is called a secondary plot.

Enzymes with ping–pong mechanisms include some oxidoreductases such as thioredoxin peroxidase, transferases such as acylneuraminate cytidylyltransferase and serine proteases such as trypsin and chymotrypsin. Serine proteases are a very common and diverse family of enzymes, including digestive enzymes (trypsin, chymotrypsin, and elastase), several enzymes of the blood clotting cascade and many others. In these serine proteases, the E* intermediate is an acyl-enzyme species formed by the attack of an active site serine residue on a peptide bond in a protein substrate. A short animation showing the mechanism of chymotrypsin is linked here.[δ]

NON-MICHAELIS–MENTEN KINETICS

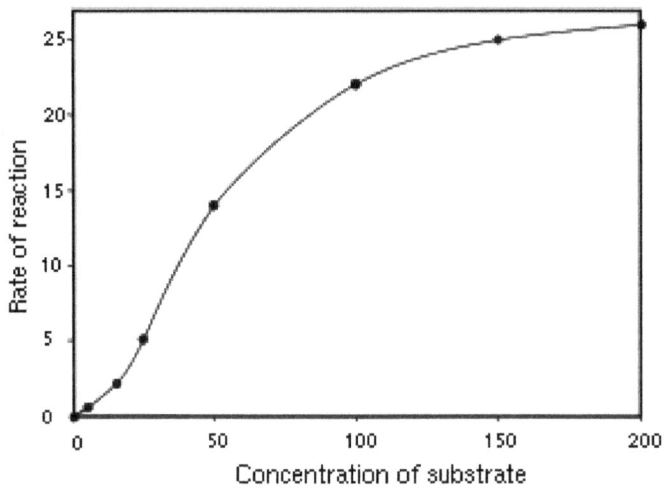

Saturation curve for an enzyme reaction showing sigmoid kinetics.

Some enzymes produce a sigmoid v by [S] plot, which often indicates cooperative binding of substrate to the active site. This means that the binding of one substrate molecule affects the binding of subsequent substrate molecules. This behavior is most common in multimeric enzymes with several interacting active sites. Here, the mechanism of cooperation is similar to that of hemoglobin, with binding of substrate to one active site altering the affinity of the other active sites for substrate molecules. Positive cooperativity occurs when binding of the first substrate molecule increases the affinity of the other active sites for substrate. Negative cooperativity occurs when binding of the first substrate decreases the affinity of the enzyme for other substrate molecules.

Allosteric enzymes include mammalian tyrosyl tRNA-synthetase, which shows negative cooperativity,and bacterial aspartate transcarbamoylase and phosphofructokinase, which show positive cooperativity.

Cooperativity is surprisingly common and can help regulate the responses of enzymes to changes in the concentrations of their substrates. Positive cooperativity makes enzymes much more sensitive to [S] and their activities can show large changes over a narrow range of substrate concentration. Conversely, negative cooperativity makes enzymes insensitive to small changes in [S].

The Hill equation (biochemistry) is often used to describe the degree of cooperativity quantitatively in non-Michaelis–Menten kinetics. The derived Hill coefficient n measures how much the binding of substrate to one active site affects the binding of substrate to the other active sites. A Hill coefficient of <1 indicates negative cooperativity and a coefficient of >1 indicates positive cooperativity.

Pre-steady-state Kinetics

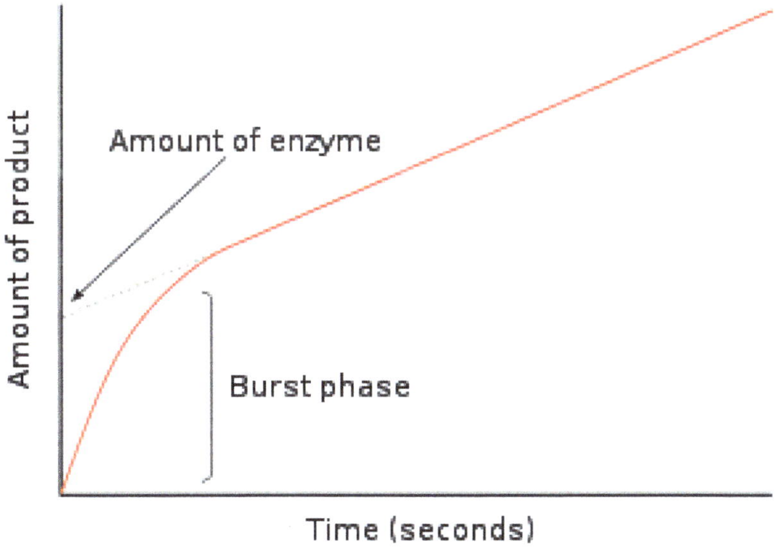

Pre-steady state progress curve, showing the burst phase of an enzyme reaction.

In the first moment after an enzyme is mixed with substrate, no product has been formed and no intermediates exist. The study of the next few milliseconds of the reaction is called pre-steady-state kinetics. Pre-steady-state kinetics is therefore concerned with the formation and consumption of enzyme–substrate intermediates (such as ES or E*) until their steady-state concentrations are reached.

This approach was first applied to the hydrolysis reaction catalysed by chymotrypsin. Often, the detection of an intermediate is a vital piece of evidence in investigations of what mechanism an enzyme follows. For example, in the ping–pong mechanisms that are shown above, rapid kinetic measurements can follow the release of product P and measure the formation of the modified enzyme intermediate E*. In the case of chymotrypsin, this intermediate is formed by an attack on the substrate by the nucleophilic serine in the active site and the formation of the acyl-enzyme intermediate.

In the figure to the right, the enzyme produces E* rapidly in the first few seconds of the reaction. The rate then slows as steady state is reached. This rapid burst phase of the reaction measures a single turnover of the enzyme. Consequently, the amount of product released in this burst, shown as the intercept on the y-axis of the graph, also gives the amount of functional enzyme which is present in the assay.

CHEMICAL MECHANISM

An important goal of measuring enzyme kinetics is to determine the chemical mechanism of an enzyme reaction, i.e., the sequence of chemical steps that transform substrate into product. The kinetic approaches discussed above will show at what rates intermediates are formed and inter-converted, but they cannot identify exactly what these intermediates are.

Kinetic measurements taken under various solution conditions or on slightly modified enzymes or substrates often shed light on this chemical mechanism, as they reveal the rate-determining step or intermediates in the reaction. For example, the breaking of a covalent bond to a hydrogen atom is a common rate-determining step. Which of the possible hydrogen transfers is rate determining can be shown by measuring the kinetic effects of substituting each hydrogen by deuterium, its stable isotope. The rate will change when the critical hydrogen is replaced, due to a primary kinetic isotope effect, which occurs because bonds to deuterium are harder to break than bonds to hydrogen. It is also possible to measure similar effects with other isotope substitutions, such as 13C/12C and 18O/16O, but these effects are more subtle.

Isotopes can also be used to reveal the fate of various parts of the substrate molecules in the final products. For example, it is sometimes difficult to discern the origin of an oxygen atom in the final product; since it may have come from water or from part of the substrate. This may be determined by systematically substituting oxygen's stable isotope 18O into the various molecules that participate in the reaction and checking for the isotope in the product. The chemical mechanism can also be elucidated by examining the kinetics and isotope effects under different pH conditions, by altering the metal ions or other bound cofactors, by site-directed mutagenesis of conserved amino acid residues, or by studying the behaviour of the enzyme in the presence of analogues of the substrate(s).

ENZYME INHIBITION AND ACTIVATION

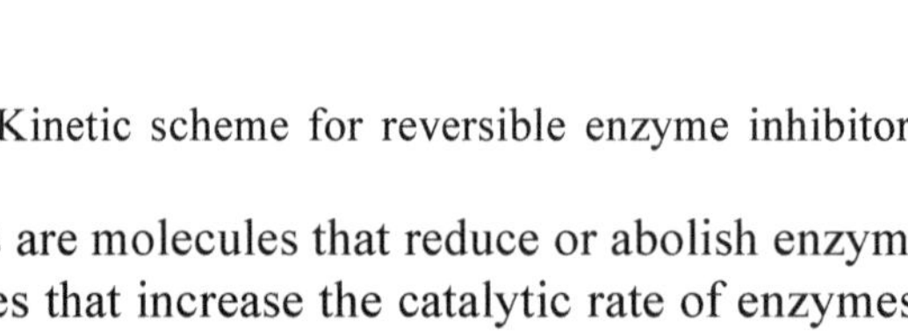

Kinetic scheme for reversible enzyme inhibitors.

Enzyme inhibitors are molecules that reduce or abolish enzyme activity, while enzyme activators are molecules that increase the catalytic rate of enzymes. These interactions can be either reversible (i.e., removal of the inhibitor restores enzyme activity) or irreversible (i.e., the inhibitor permanently inactivates the enzyme).

Reversible Inhibitors

Traditionally reversible enzyme inhibitors have been classified as competitive, uncompetitive, or non-competitive, according to their effects on Km and Vmax. These different effects result from the inhibitor binding to the enzyme E, to the enzyme–substrate complex ES, or to both, respectively. The division of these classes arises from a problem in their derivation and results in the need to use two different binding constants for one binding event. The binding of an inhibitor and its effect on the enzymatic activity are two distinctly different things, another problem the traditional equations fail to acknowledge. In noncompetitive inhibition the binding of the inhibitor results in 100% inhibition of the enzyme only, and fails to consider the possibility of anything in between. The common form of the inhibitory term also obscures the relationship between the inhibitor binding to the enzyme and its relationship to any other binding term be it the Michaelis–Menten equation or a dose response curve associated with ligand receptor binding. To demonstrate the relationship the following rearrangement can be made:

$$\frac{V_{\max}}{1+\frac{[I]}{K_i}} = \frac{V_{\max}}{\frac{[I]+K_i}{K_i}}$$

Adding zero to the bottom ([I]-[I])

$$\frac{V_{\max}}{\frac{[I]+K_i}{[I]+K_i-[I]}}$$

Dividing by $[I]+K_i$

$$\frac{V_{\max}}{\frac{1}{1-\frac{[I]}{[I]+K_i}}} = V_{\max} - V_{\max}\frac{[I]}{[I]+K_i}$$

This notation demonstrates that similar to the Michaelis–Menten equation, where the rate of reaction depends on the percent of the enzyme population interacting with substratefraction of the enzyme population bound by substrate

$$\frac{[S]}{[S]+K_m}$$

fraction of the enzyme population bound by inhibitor

$$\frac{[I]}{[I]+K_i}$$

the effect of the inhibitor is a result of the percent of the enzyme population interacting with inhibitor. The only problem with this equation in its present form is that it assumes absolute inhibition of the enzyme with inhibitor binding, when in fact there can be a wide range of effects anywhere from 100% inhibition of substrate turn over to just >0%. To account for this the equation can be easily modified to allow for different degrees of inhibition by including a delta Vmax term.

$$V_{\max} - \Delta V_{\max}\frac{[I]}{[I]+K_i}$$

or

$$V_{\max 1} - (V_{\max 1} - V_{\max 2})\frac{[I]}{[I]+K_i}$$

This term can then define the residual enzymatic activity present when the inhibitor is interacting with individual enzymes in the population. However the inclusion of this term has the added value of allowing for the possibility of activation if the secondary Vmax term turns out to be higher than the initial term. To account for the possibly of activation as well the notation can then be rewritten replacing the inhibitor "I" with a modifier term denoted here as "X".

$$V_{\max 1} - (V_{\max 1} - V_{\max 2})\frac{[X]}{[X]+K_x}$$

While this terminology results in a simplified way of dealing with kinetic effects relating to the maximum velocity of the Michaelis–Menten equation, it highlights potential problems with the term used to describe effects relating to the Km. The Km relating to the affinity of the enzyme for the substrate should in most cases relate to potential changes in the binding site of the enzyme which would directly result from enzyme inhibitor interactions. As such a term similar to the one proposed above to modulate Vmaxshould be appropriate in most situations:

$$K_{m1} - (K_{m1} - K_{m2})\frac{[X]}{[X]+K_x}$$

Irreversible Inhibitors

Enzyme inhibitors can also irreversibly inactivate enzymes, usually by covalently modifying active site residues. These reactions, which may be called suicide substrates, follow exponential decay functions and are usually saturable. Below saturation, they follow first order kinetics with respect to inhibitor.

MECHANISMS OF CATALYSIS

The favoured model for the enzyme–substrate interaction is the induced fit model. This model proposes that the initial interaction between enzyme and substrate is relatively weak, but that these weak interactions rapidly induce conformational changes in the enzyme that strengthen binding. These conformational changes also bring catalytic residues in the active site close to the chemical bonds in the substrate that will be altered in the reaction. Conformational changes can be measured using circular dichroism or dual polarisation interferometry. After binding takes place, one or more mechanisms of catalysis lower the energy of the reaction's transition state by providing an alternative chemical pathway for the reaction. Mechanisms of catalysis include catalysis by bond strain; by proximity and orientation; by active-site proton donors or acceptors; covalent catalysis and quantum tunnelling.

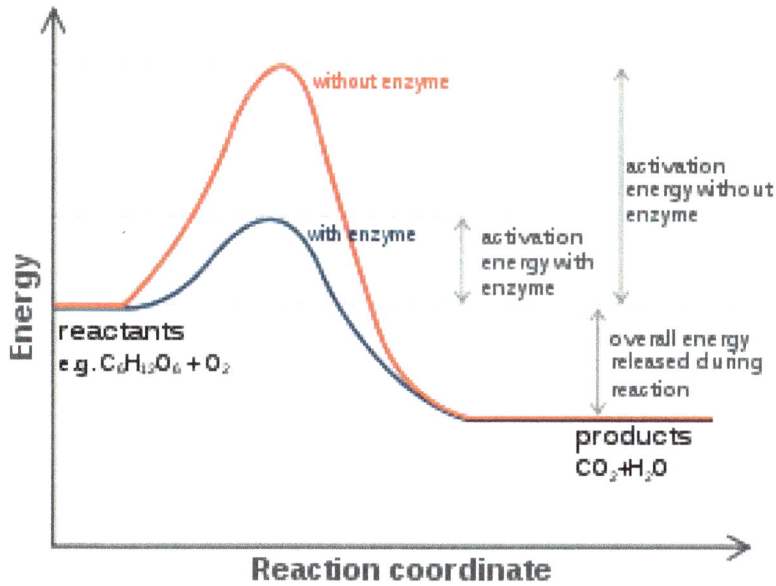

The energy variation as a function of reaction coordinate shows the stabilisation of the transition state by an enzyme.

Enzyme kinetics cannot prove which modes of catalysis are used by an enzyme. However, some kinetic data can suggest possibilities to be examined by other techniques. For example, a ping–pong mechanism with burst-phase pre-steady-state kinetics would suggest covalent catalysis might be important in this enzyme's mechanism. Alternatively, the observation of a strong pH effect on Vmax but not Km might indicate that a residue in the active site needs to be in a particular ionisation state for catalysis to occur.

HISTORY

In 1902 Victor Henri proposed a quantitative theory of enzyme kinetics, but at the time the experimental significance of the hydrogen ion concentration was not yet recognized. After Peter Lauritz Sørensen had defined the logarithmic pH-scale and introduced the concept of buffering in 1909 the German chemist Leonor Michaelis and his Canadian postdoc Maud Leonora Menten repeated Henri's experiments and confirmed his equation, which is now generally referred to as Michaelis-Menten kinetics (sometimes also Henri-Michaelis-Menten kinetics). Their work was further developed by G. E. Briggs and J. B. S. Haldane, who derived kinetic equations that are still widely considered today a starting point in modeling enzymatic activity.

The major contribution of the Henri-Michaelis-Menten approach was to think of enzyme reactions in two stages. In the first, the substrate binds reversibly to the enzyme, forming the enzyme-substrate complex. This is sometimes called the Michaelis complex. The enzyme then catalyzes the chemical step in the reaction and releases the product. The kinetics of many enzymes is adequately described by the simple Michaelis-Menten model, but all enzymes have internal motions that are not accounted for in the model and can have significant contributions to the overall reaction kinetics. This can be modeled by introducing several Michaelis-Menten pathways that are connected with fluctuating rates, which is a mathematical extension of the basic Michaelis Menten mechanism.

SOFTWARE

ENZO

ENZO (Enzyme Kinetics) is a graphical interface tool for building kinetic models of enzyme catalyzed reactions. ENZO automatically generates the corresponding differential equations from a stipulated enzyme reaction scheme. These differential equations are processed by a numerical solver and a regression algorithm which fits the coefficients of differential equations to experimentally observed time course curves. ENZO allows rapid evaluation of rival reaction schemes and can be used for routine tests in enzyme kinetics.

EADIE–HOFSTEE DIAGRAM

In biochemistry, an Eadie–Hofstee diagram (also Woolf–Eadie–Augustinsson–Hofstee or Eadie–Augustinsson plot) is a graphical representation of enzyme kinetics in which reaction rate is plotted as a function of the ratio between rate and substrate concentration:

$$v = -K_m \frac{v}{[S]} + V_{\max}$$

where v represents reaction rate, Km is the Michaelis–Menten constant, [S] is the substrate concentration, and Vmax is the maximum reaction rate.

It can be derived from the Michaelis–Menten equation as follows:

$$v = \frac{V_{\max}[S]}{K_m + [S]}$$

invert and multiply with $V_{\max}$:

$$\frac{V_{\max}}{v} = \frac{V_{\max}(K_m + [S])}{V_{\max}[S]} = \frac{K_m + [S]}{[S]}$$

Rearrange:

$$V_{\max} = \frac{vK_m}{[S]} + \frac{v[S]}{[S]} = \frac{vK_m}{[S]} + v$$

Isolate v:

$$v = -K_m \frac{v}{[S]} + V_{\max}$$

A plot of v against v/[S] will hence yield Vmax as the y-intercept, Vmax/Km as the x-intercept, and Km as the negative slope. Like other techniques that linearize the Michaelis–Menten equation, the Eadie-Hofstee plot was used historically for rapid identification of important kinetic terms like Km and Vmax, but has been superseded by nonlinear regression methods that are significantly more accurate and no longer computationally inaccessible. It is also more robust against error-prone data than the Lineweaver–Burk plot, particularly because it gives equal weight to data points in any range of substrate concentration or reaction rate. (The Lineweaver–Burk plot unevenly weights such points.) Both plots remain useful as a means to present data graphically.

One drawback from the Eadie–Hofstee approach is that neither ordinate nor abscissa represent independent variables: both are dependent on reaction rate. Thus any experimental error will be present in both axes. Also, experimental error or uncertainty will propagate unevenly and become larger over the abscissa thereby giving more weight to smaller values of v/[S]. Therefore, the typical measure of goodness of fit for linear regression, the correlation coefficient R, is not applicable.

HANES–WOOLF PLOT

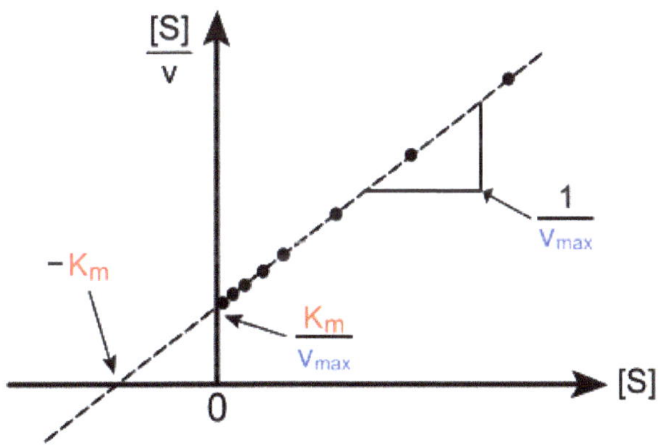

Hanes–Woolf plot.

In biochemistry, a Hanes–Woolf plot is a graphical representation of enzyme kinetics in which the ratio of the initial substrate concentration [S] to the reaction velocity v is plotted against [S]. It is based on the rearrangement of the Michaelis–Menten equation shown below:

$$\frac{[S]}{v} = \frac{[S]}{V_{\max}} + \frac{K_m}{V_{\max}}$$

where Km is the Michaelis–Menten constant and Vmax is the maximum reaction velocity.

J B S Haldane stated that this method was due to Barnet Woolf. It was also used by Charles Samuel Hanes, even though he neither mentions nor cites Woolf. Hanes pointed out that the use of linear regression to determine kinetic parameters from this type of linear transformation is flawed, because it generates the best fit between observed and calculated values of 1/v, rather than v.

The equation can be derived from the Michaelis–Menten equation as follows:

$$v = \frac{V_{\max}[S]}{K_m + [S]}$$

invert and multiply by [S]:

$$\frac{[S]}{v} = \frac{[S](K_m + [S])}{V_{\max}[S]} = \frac{K_m + [S]}{V_{\max}}$$

Rearrange:

$$\frac{[S]}{v} = \frac{1}{V_{\max}}[S] + \frac{K_m}{V_{\max}}$$

As is clear from the equation, perfect data will yield a straight line of slope 1/Vmax, a y-intercept of Km/Vmax and an x-intercept of −Km.

Like other techniques that linearize the Michaelis–Menten equation, the Hanes–Woolf plot was used historically for rapid determination of the important kinetic parameters Km, Vmax and Vmax/Km, but it has been superseded by nonlinear regression methods that are significantly more accurate and no longer computationally inaccessible. It remains useful, however, as a means to present data graphically.

One drawback of the Hanes–Woolf approach is that neither ordinate nor abscissa represent independent variables: both are dependent on substrate concentration. As a result, the typical measure of goodness of fit, the correlation coefficient R, is not applicable.

LINEWEAVER–BURK PLOT

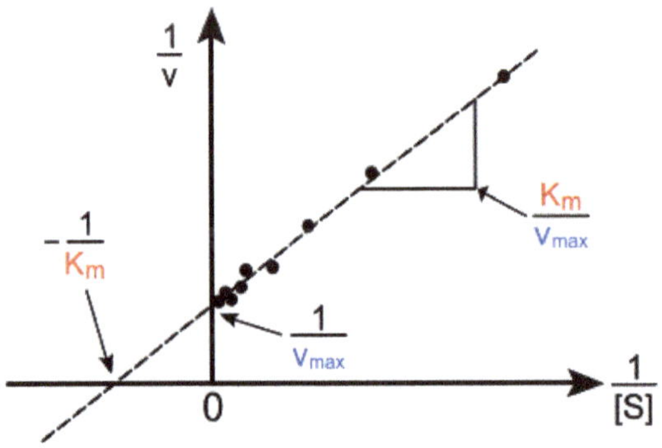

An example of a Lineweaver-Burk plot.

In biochemistry, the Lineweaver–Burk plot (or double reciprocal plot) is a graphical representation of the Lineweaver–Burk equation of enzyme kinetics, described by Hans Lineweaver and Dean Burk in 1934.

Derivation

The plot provides a useful graphical method for analysis of the Michaelis–Menten equation:

$$V = \frac{V_{\max}[S]}{K_m + [S]}$$

Taking the reciprocal gives

$$\frac{1}{V} = \frac{K_m + [S]}{V_{\max}[S]} = \frac{K_m}{V_{\max}}\frac{1}{[S]} + \frac{1}{V_{\max}}$$

where V is the reaction velocity (the reaction rate), Km is the Michaelis–Menten constant, Vmax is the maximum reaction velocity, and [S] is the substrate concentration.

Use

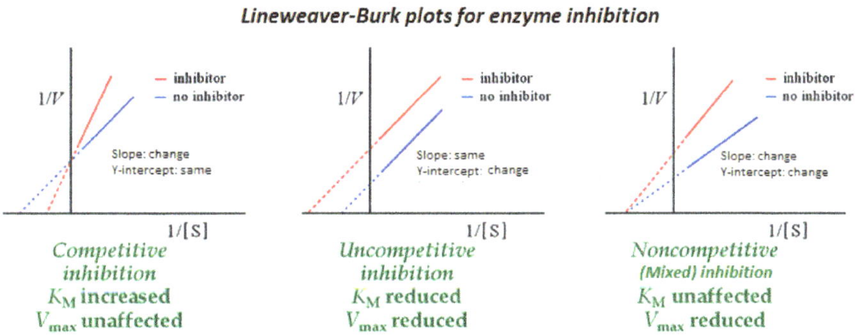

Enzyme Inhibition displayed using Lineweaver-Burk (double reciprocal plots).

The Lineweaver–Burk plot was widely used to determine important terms in enzyme kinetics, such as Km and Vmax, before the wide availability of powerful computers and non-linear regression software. The y-intercept of such a graph is equivalent to the inverse of Vmax; the x-intercept of the graph represents −1/Km. It also gives a quick, visual impression of the different forms of enzyme inhibition.

The double reciprocal plot distorts the error structure of the data, and it is therefore unreliable for the determination of enzyme kinetic parameters. Although it is still used for representation of kinetic data,non-linear regression or alternative linear forms of the Michaelis–Menten equation such as the Hanes-Woolf plot or Eadie–Hofstee plot are generally used for the calculation of parameters.

When used for determining the type of enzyme inhibition, the Lineweaver–Burk plot can distinguish competitive, non-competitive and uncompetitive inhibitors. Competitive inhibitors have the same y-intercept as uninhibited enzyme (since Vmax is unaffected by competitive inhibitors the inverse of Vmaxalso doesn't change) but there are different slopes and x-intercepts between the two data sets. Non-competitive inhibition produces plots with the same x-intercept as uninhibited enzyme (Km is unaffected) but different slopes and y-intercepts. Uncompetitive inhibition causes different intercepts on both the y- and x-axes .

Problems with the Method

The Lineweaver–Burk plot is classically used in older texts, but is prone to error, as the y-axis takes the reciprocal of the rate of reaction – in turn increasing any small errors in measurement. Also, most points on the plot are found far to the right of the y-axis (due to limiting solubility not allowing for large values of [S] and hence no small values for 1/[S]), calling for a large extrapolation back to obtain x- and y-intercepts.

REFERENCES

1. http://www.worthington-biochem.com/introbiochem/substrateconc.html
2. Henri, Victor (1903). Lois Générales de l'Action des Diastases. Paris: Hermann. Google books (US only)
3. "Victor Henri". Whonamedit?. Retrieved 24 May 2011.
4. Michaelis, L.; Menten, M.L. (1913). "Die Kinetik der Invertinwirkung". Biochem Z. 49: 333–369 (recent translation, and an older partial translation)
5. Chen, W.W.; Neipel, M.; Sorger, P.K. (2010). "Classic and contemporary approaches to modeling biochemical reactions". Genes Dev. 24 (17): 1861–1875. doi:10.1101/gad.1945410. PMC 2932968. PMID 20810646.
6. Laidler K.J. and Meiser J.H. Physical Chemistry (Benjamin/Cummings 1982) p.430 ISBN 0-8053-5682-7
7. Lehninger, A.L.; Nelson, D.L.; Cox, M.M. (2005). Lehninger principles of biochemistry. New York: W.H. Freeman. ISBN 978-0-7167-4339-2.
8. Chakraborty, S. (23 Dec 2009). Microfluidics and Microfabrication (1 ed.). Springer. ISBN 978-1-4419-1542-9.
9. Mathews, C.K.; van Holde, K.E.; Ahern, K.G. (10 Dec 1999). Biochemistry (3 ed.). Prentice Hall. ISBN 978-0-8053-3066-3.
10. Stroppolo, M.E.; Falconi, M.; Caccuri, A.M.; Desideri, A. (Sep 2001). "Superefficient enzymes". Cell Mol Life Sci. 58 (10): 1451–60. doi:10.1007/PL00000788. PMID 11693526.
11. Yu, R.C.; Rappaport, S.M. (1997). "A lung retention model based on Michaelis–Menten-like kinetics". Environ Health Perspect. 105 (5): 496–503. doi:10.1289/

ehp.97105496. PMC 1469867 PMID 9222134.

12. Keating, K.A.; Quinn, J.F. (1998). "Estimating species richness: the Michaelis–Menten model revisited". Oikos. 81 (2): 411–416. doi:10.2307/3547060. JSTOR 3547060.

13. Jones, A.W. (2010). "Evidence-based survey of the elimination rates of ethanol from blood with applications in forensic casework". Forensic Sci Int. 200 (1–3): 1–20. doi:10.1016/j.forsciint.2010.02.021. PMID 20304569.

14. Abedon, S.T. (2009). "Kinetics of phage-mediated biocontrol of bacteria". Foodborne Pathog Dis. 6 (7): 807–15. doi:10.1089/fpd.2008.0242. PMID 19459758.

15. Murray, J.D. (2002). Mathematical Biology: I. An Introduction (3 ed.). Springer. ISBN 978-0-387-95223-9.

16. Keener, J.; Sneyd, J. (2008). Mathematical Physiology: I: Cellular Physiology (2 ed.). Springer. ISBN 978-0-387-75846-6.

17. Briggs, G.E.; Haldane, J.B.S. (1925). "A note on the kinematics of enzyme action". Biochem J. 19 (2): 338–339. doi:10.1042/bj0190338. PMC 1259181. PMID 16743508.

18. Zhou, H.X.; Rivas, G.; Minton, A.P. (2008). "Macromolecular crowding and confinement: biochemical, biophysical, and potential physiological consequences". Annu Rev Biophys. 37 (1): 375–97. doi:10.1146/annurev.biophys.37.032807.125817. PMC 2826134. PMID 18573087.

19. Grima, R.; Schnell, S. (Oct 2006). "A systematic investigation of the rate laws valid in intracellular environments". Biophys Chem. 124 (1): 1–10. doi:10.1016/j.bpc.2006.04.019. PMID 16781049.

20. Schnell, S.; Turner, T.E. (2004). "Reaction kinetics in intracellular environments with macromolecular crowding: simulations and rate laws". Prog Biophys Mol Biol. 85 (2–3): 235–60. doi:10.1016/j.pbiomolbio.2004.01.012. PMID 15142746.

21. Segel, L.A.; Slemrod, M. (1989). "The quasi-steady-state assumption: A case study in perturbation". Thermochim Acta. 31 (3): 446–477. doi:10.1137/1031091.

22. Leskovac, V. (2003). Comprehensive enzyme kinetics. New York: Kluwer Academic/Plenum Pub. ISBN 978-0-306-46712-7.

23. Greco, W.R.; Hakala, M.T. (1979). "Evaluation of methods for estimating the dissociation constant of tight binding enzyme inhibitors,". J Biol Chem. 254 (23): 12104–12109. PMID 500698.

24. Hayakawa, K.; Guo, L.; Terentyeva, E.A.; Li, X.K.; Kimura, H.; Hirano, M.; Yoshikawa, K.; Nagamine, T.; et al. (2006). "Determination of specific activities and kinetic constants of biotinidase and lipoamidase in LEW rat and Lactobacillus casei (Shirota)". J Chromatogr B Analyt Technol Biomed Life Sci. 844 (2): 240–50. doi:10.1016/j.jchromb.2006.07.006. PMID 16876490.

25. Schnell, S.; Mendoza, C. (1997). "A closed form solution for time-dependent enzyme kinetics". Journal of Theoretical Biology. 187 (2): 207–212. doi:10.1006/

jtbi.1997.0425.

26. Goudar, C. T.; Sonnad, J. R.; Duggleby, R. G. (1999). "Parameter estimation using a direct solution of the integrated Michaelis–Menten equation". Biochimica et Biophysica Acta (BBA) - Protein Structure and Molecular Enzymology. 1429 (2): 377–383. doi:10.1016/s0167-4838(98)00247-7. PMID 9989222.

27. Goudar, C. T.; Harris, S. K.; McInerney, M. J.; Suflita, J. M. (2004). "Progress curve analysis for enzyme and microbial kinetic reactions using explicit solutions based on the Lambert W function". Journal of Microbiological Methods. 59 (3): 317–326. doi:10.1016/j.mimet.2004.06.013. PMID 15488275.

28. Reuveni, Shlomi; Urbakh, Michael; Klafter, Joseph (2014). "Role of Substrate Unbinding in Michaelis-Menten Enzymatic Reactions". Proceedings of the National Academy of Sciences. 111 (12): 4391–4396. Bibcode:2014PNAS..111.4391R. doi:10.1073/pnas.1318122111.

29. Wrighton, Mark S.; Ebbing, Darrell D. (1993). General chemistry (4th ed.). Boston: Houghton Mifflin. ISBN 0-395-63696-5.

30. Danson, Michael; Eisenthal, Robert (2002). Enzyme assays: a practical approach. Oxford [Oxfordshire]: Oxford University Press. ISBN 0-19-963820-9.

31. Xie XS, Lu HP (June 1999). "Single-molecule enzymology". J. Biol. Chem. ***274*** (23): 15967–70. doi:10.1074/jbc.274.23.15967. PMID 10347141.

32. Lu H (2004). "Single-molecule spectroscopy studies of conformational change dynamics in enzymatic reactions". Current pharmaceutical biotechnology. ***5*** (3): 261–9. doi:10.2174/1389201043376887. PMID 15180547.

33. Schnell J, Dyson H, Wright P (2004). "Structure, dynamics, and catalytic function of dihydrofolate reductase". Annual Review of Biophysics and Biomolecular Structure. ***33***: 119–40. doi:10.1146/annurev.biophys.33.110502.133613. PMID 15139807.

34. Gibson QH (1969). "Rapid mixing: Stopped flow". Methods in Enzymology. Methods in Enzymology. ***16***: 187–228. doi:10.1016/S0076-6879(69)16009-7. ISBN 978-0-12-181873-9.

35. Duggleby RG (1995). "Analysis of enzyme progress curves by non-linear regression". Methods in Enzymology. Methods in Enzymology. ***249***: 61–90. doi:10.1016/0076-6879(95)49031-0. ISBN 978-0-12-182150-0. PMID 7791628.

36. Murray JB, Dunham CM, Scott WG (January 2002). "A pH-dependent conformational change, rather than the chemical step, appears to be rate-limiting in the hammerhead ribozyme cleavage reaction". J. Mol. Biol. ***315*** (2): 121–30. doi:10.1006/jmbi.2001.5145. PMID 11779233.

37. Michaelis L. and Menten M.L. Kinetik der Invertinwirkung Biochem. Z. 1913; 49:333–369 English translation Accessed 6 April 2007

38. Stroppolo ME, Falconi M, Caccuri AM, Desideri A (2001). "Superefficient enzymes". Cell. Mol. Life Sci. ***58*** (10): 1451–60. doi:10.1007/PL00000788. PMID 11693526.

39. Walsh R, Martin E, Darvesh S (Jan 2010). "A method to describe enzyme-catalyzed

reactions by combining steady state and time course enzyme kinetic parameters". Biochim Biophys Acta. ***1800*** (1): 1–5. doi:10.1016/j.bbagen.2009.10.007. PMID 19840832.

40. Beal, S. L. (1983). "Computation of the explicit solution to the Michaelis-Menten equation". Journal of Pharmacokinetics and Biopharmaceutics. ***11*** (6): 641–657. doi:10.1007/BF01059062. PMID 6689584.
41. Schnell, S.; Mendoza, C. (1997). "Closed Form Solution for Time-dependent Enzyme Kinetics". Journal of Theoretical Biology. ***187*** (2): 207–212. doi:10.1006/jtbi.1997.0425.
42. Goudar, CT; Sonnad, JR; Duggleby, RG (1999). "Parameter estimation using a direct solution of the integrated Michaelis-Menten equation" (PDF). Biochimica et Biophysica Acta. ***1429*** (2): 377–83. doi:10.1016/s0167-4838(98)00247-7. PMID 9989222. Archived from the original (PDF) on 9 November 2015.
43. Goudar, C. T.; Harris, S. K.; McInerney, M. J.; Suflita, J. M. (2004). "Progress curve analysis for enzyme and microbial kinetic reactions using explicit solutions based on the Lambert W function". Journal of Microbiological Methods. ***59*** (3): 317–326. doi:10.1016/j.mimet.2004.06.013. PMID 15488275.
44. Jones ME (1992). "Analysis of algebraic weighted least-squares estimators for enzyme parameters". Biochem. J. ***288*** (Pt 2): 533–8. PMC 1132043. PMID 1463456.
45. Tseng SJ, Hsu JP (August 1990). "A comparison of the parameter estimating procedures for the Michaelis-Menten model". J. Theor. Biol. ***145*** (4): 457–64. doi:10.1016/S0022-5193(05)80481-3. PMID 2246896.
46. Bravo IG, Busto F, De Arriaga D, et al. (September 2001). "A normalized plot as a novel and time-saving tool in complex enzyme kinetic analysis". Biochem. J. ***358*** (Pt 3): 573–83. PMC 1222113. PMID 11577687.
47. Almaas E, Kovács B, Vicsek T, Oltvai ZN, Barabási AL (February 2004). "Global organization of metabolic fluxes in the bacterium Escherichia coli". Nature. ***427*** (6977): 839–43. Bibcode:2004Natur.427..839A. doi:10.1038/nature02289. PMID 14985762.
48. Reed JL, Vo TD, Schilling CH, Palsson BO (2003). "An expanded genome-scale model of Escherichia coli K-12 (iJR904 GSM/GPR)". Genome Biol. ***4*** (9): R54. doi:10.1186/gb-2003-4-9-r54. PMC 193654. PMID 12952533.
49. for a complete derivation, see here
50. Dirr H, Reinemer P, Huber R (1994). "X-ray crystal structures of cytosolic glutathione S-transferases. Implications for protein architecture, substrate recognition and catalytic function". Eur. J. Biochem. ***220*** (3): 645–61. doi:10.1111/j.1432-1033.1994.tb18666.x. PMID 8143720.
51. Stone SR, Morrison JF (July 1988). "Dihydrofolate reductase from Escherichia coli: the kinetic mechanism with NADPH and reduced acetylpyridine adenine dinucleotide phosphate as substrates". Biochemistry. ***27*** (15): 5493–9. doi:10.1021/bi00415a016. PMID 3052577.
52. Fisher PA (1994). "Enzymologic mechanism of replicative DNA polymerases in higher eukaryotes". Progress in Nucleic Acid Research and Molecular Biology.

Progress in Nucleic Acid Research and Molecular Biology. ***47***: 371–97. doi:10.1016/S0079-6603(08)60257-3. ISBN 978-0-12-540047-3. PMID 8016325.

53. Akerman SE, Müller S (2003). "2-Cys peroxiredoxin PfTrx-Px1 is involved in the antioxidant defence of Plasmodium falciparum". Mol. Biochem. Parasitol. ***130*** (2): 75–81. doi:10.1016/S0166-6851(03)00161-0. PMID 12946843.

54. Bravo IG, Barrallo S, Ferrero MA, Rodríguez-Aparicio LB, Martínez-Blanco H, Reglero A (September 2001). "Kinetic properties of the acylneuraminate cytidylyltransferase from Pasteurella haemolytica A2". Biochem. J. ***358*** (Pt 3): 585–98. PMC 1222114. PMID 11577688.

55. Kraut J (1977). "Serine proteases: structure and mechanism of catalysis". Annu. Rev. Biochem. ***46***: 331–58. doi:10.1146/annurev.bi.46.070177.001555. PMID 332063.

56. Ricard J, Cornish-Bowden A (July 1987). "Co-operative and allosteric enzymes: 20 years on". Eur. J. Biochem. ***166*** (2): 255–72. doi:10.1111/j.1432-1033.1987.tb13510.x. PMID 3301336.

57. Ward WH, Fersht AR (July 1988). "Tyrosyl-tRNA synthetase acts as an asymmetric dimer in charging tRNA. A rationale for half-of-the-sites activity". Biochemistry. ***27*** (15): 5525–30. doi:10.1021/bi00415a021. PMID 3179266.

58. Helmstaedt K, Krappmann S, Braus GH (September 2001). "Allosteric Regulation of Catalytic Activity: Escherichia coli Aspartate Transcarbamoylase versus Yeast Chorismate Mutase". Microbiol. Mol. Biol. Rev. ***65*** (3): 404–21, table of contents. doi:10.1128/MMBR.65.3.404-421.2001. PMC 99034. PMID 11528003.

59. Schirmer T, Evans PR (January 1990). "Structural basis of the allosteric behaviour of phosphofructokinase". Nature. ***343*** (6254): 140–5. Bibcode:1990Natur.343..140S. doi:10.1038/343140a0. PMID 2136935.

60. Hill, A. V. The possible effects of the aggregation of the molecules of haemoglobin on its dissociation curves. J. Physiol. (Lond.), 1910 40, iv–vii.

61. Hartley BS, Kilby BA (February 1954). "The reaction of p-nitrophenyl esters with chymotrypsin and insulin". Biochem. J. ***56*** (2): 288–97. PMC 1269615. PMID 13140189.

62. Fersht, Alan (1999). Structure and mechanism in protein science: a guide to enzyme catalysis and protein folding. San Francisco: W.H. Freeman. ISBN 0-7167-3268-8.

63. Bender ML, Begué-Cantón ML, Blakeley RL, et al. (December 1966). "The determination of the concentration of hydrolytic enzyme solutions: alpha-chymotrypsin, trypsin, papain, elastase, subtilisin, and acetylcholinesterase". J. Am. Chem. Soc. ***88*** (24): 5890–913. doi:10.1021/ja00976a034. PMID 5980876.

64. Cleland WW (January 2005). "The use of isotope effects to determine enzyme mechanisms". Arch. Biochem. Biophys. ***433*** (1): 2–12. doi:10.1016/j.abb.2004.08.027. PMID 15581561.

65. Northrop D (1981). "The expression of isotope effects on enzyme-catalyzed reactions". Annu. Rev. Biochem. ***50***: 103–31. doi:10.1146/annurev.bi.50.070181.000535. PMID 7023356.

66. Baillie T, Rettenmeier A (1986). "Drug biotransformation: mechanistic studies with stable isotopes". Journal of clinical pharmacology. ***26*** (6): 448–51. doi:10.1002/j.1552-4604.1986.tb03556.x. PMID 3734135.

67. Cleland WW (1982). "Use of isotope effects to elucidate enzyme mechanisms". CRC Crit. Rev. Biochem. ***13*** (4): 385–428. doi:10.3109/10409238209108715. PMID 6759038.

68. Christianson DW, Cox JD (1999). "Catalysis by metal-activated hydroxide in zinc and manganese metalloenzymes". Annu. Rev. Biochem. ***68***: 33–57. doi:10.1146/annurev.biochem.68.1.33. PMID 10872443.

69. Kraut D, Carroll K, Herschlag D (2003). "Challenges in enzyme mechanism and energetics". Annu. Rev. Biochem. ***72***: 517–71. doi:10.1146/annurev.biochem.72.121801.161617. PMID 12704087.

70. Walsh, R.; Martin, E.; Darvesh, S. (2011). "Limitations of conventional inhibitor classifications". Integrative Biology. Royal Society of Chemistry. ***3*** (12): 1197–1201. doi:10.1039/c1ib00053e. PMID 22038120.

71. Walsh, R.; Martin, E.; Darvesh, S. (2007). "A versatile equation to describe reversible enzyme inhibition and activation kinetics: Modeling β-galactosidase and butyrylcholinesterase". Biochimica et Biophysica Acta (BBA) - General Subjects. ***1770*** (5): 733–746. doi:10.1016/j.bbagen.2007.01.001. PMID 17307293.

72. Walsh, Ryan (2012). "Ch. 17. Alternative Perspectives of Enzyme Kinetic Modeling". In Ekinci, Deniz. Medicinal Chemistry and Drug Design (PDF). InTech. pp. 357–371. ISBN 978-953-51-0513-8.

73. Koshland DE (February 1958). "Application of a Theory of Enzyme Specificity to Protein Synthesis". Proc. Natl. Acad. Sci. U.S.A. ***44*** (2): 98–104. Bibcode:1958PNAS...44...98K. doi:10.1073/pnas.44.2.98. PMC 335371. PMID 16590179.

74. Hammes G (2002). "Multiple conformational changes in enzyme catalysis". Biochemistry. ***41*** (26): 8221–8. doi:10.1021/bi0260839. PMID 12081470.

75. Sutcliffe M, Scrutton N (2002). "A new conceptual framework for enzyme catalysis. Hydrogen tunnelling coupled to enzyme dynamics in flavoprotein and quinoprotein enzymes". Eur. J. Biochem. ***269*** (13): 3096–102. doi:10.1046/j.1432-1033.2002.03020.x. PMID 12084049.

76. Henri V (1902). "Theorie generale de l'action de quelques diastases". Compt. Rend. Hebd. Acad. Sci. Paris. ***135***: 916–9.

77. Sørensen PL (1909). "Enzymstudien {II}. Über die Messung und Bedeutung der Wasserstoffionenkonzentration bei enzymatischen Prozessen" [Enzyme studies III: About the measurement and significance of the hydrogen ion concentration in enzymatic processes]. Biochem. Z. (in German). ***21***: 131–304.

78. Michaelis L, Menten M (1913). "Die Kinetik der Invertinwirkung" [The Kinetics of Invertase Action]. Biochem. Z. (in German). ***49***: 333–369.; Michaelis L, Menten ML, Johnson KA, Goody RS (2011). "The original Michaelis constant: translation

of the 1913 Michaelis-Menten paper". Biochemistry. ***50*** (39): 8264–9. doi:10.1021/bi201284u. PMC 3381512. PMID 21888353.

79. Briggs GE, Haldane JB (1925). "A Note on the Kinetics of Enzyme Action". The Biochemical Journal. ***19*** (2): 339–339. doi:10.1042/bj0190338. PMC 1259181. PMID 16743508.

80. Flomenbom O, Velonia K, Loos D, Masuo S, Cotlet M, Engelborghs Y, Hofkens J, Rowan AE, Nolte RJ, Van der Auweraer M, de Schryver FC, Klafter J (Feb 2005). "Stretched exponential decay and correlations in the catalytic activity of fluctuating single lipase molecules". Proceedings of the National Academy of Sciences of the United States of America. ***102*** (7): 2368–2372. Bibcode:2005PNAS..102.2368F. doi:10.1073/pnas.0409039102. PMC 548972. PMID 15695587.

81. English BP, Min W, van Oijen AM, Lee KT, Luo G, Sun H, Cherayil BJ, Kou SC, Xie XS (Feb 2006). "Ever-fluctuating single enzyme molecules: Michaelis-Menten equation revisited". Nature Chemical Biology. ***2*** (2): 87–94. doi:10.1038/nchembio759. PMID 16415859.

82. Lu HP, Xun L, Xie XS (Dec 1998). "Single-molecule enzymatic dynamics". Science. ***282*** (5395): 1877–1882. doi:10.1126/science.282.5395.1877. PMID 9836635.

83. Xue X, Liu F, Ou-Yang ZC (Sep 2006). "Single molecule Michaelis-Menten equation beyond quasistatic disorder". Physical Review E. ***74*** (3 Pt 1): 030902. Bibcode:2006PhRvE..74c0902X. doi:10.1103/PhysRevE.74.030902. PMID 17025584.

84. Bevc S.; Konc J.; Stojan J.; Hodoščeк M.; Penca M.; Matej Praprotnik M.; Janežič D. (2011). "ENZO: A Web Tool for Derivation and Evaluation of Kinetic Models of Enzyme Catalyzed Reactions". PLoS ONE. ***6*** (7): e22265. doi:10.1371/journal.pone.0022265. PMC 3139599. PMID 21818304.

CHAPTER 6

Allosteric Regulation

Chapter Contents

- **Introduction**
- **Allosteric Resources Online**
- **Allosteric Modulation**
- **Types**
- **Non-Regulatory Allostery**
- **Pharmacology**
- **Cooperativity**
- **Enzyme Inhibitor**

Enzymes have an area called the active site, where they bind substrates, like the hamburger, and turn them into products or food for cells. Many enzymes have other areas called allosteric sites, located in a different place from the active site.

An allosteric site does not bind substrate, but instead binds another molecule that affects the enzyme's regulation. When a molecule binds an allosteric site, it alters the enzyme's shape, or conformation, which then changes how the enzyme functions.

To understand this concept more clearly, think about your kitchen lights. How do you turn them on and off? Do you screw and unscrew the light bulbs every time you want to change the lighting? No. You flick a switch. Imagine that the allosteric site represents the light switch and that by controlling the switch, you control the light bulb, or the enzyme's active site.

INTRODUCTION

In biochemistry, allosteric regulation (or allosteric control) is the regulation of an enzyme by binding an effector molecule at a site other than the enzyme's active site.

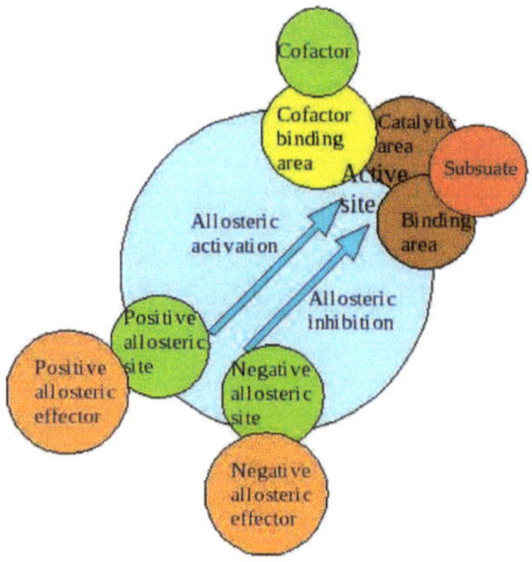

Allosteric regulation of an enzyme.

The site to which the effector binds is termed the allosteric site. Allosteric sites allow effectors to bind to the protein, often resulting in a conformational change involving protein dynamics. Effectors that enhance the protein's activity are referred to as allosteric activators, whereas those that decrease the protein's activity are called allosteric inhibitors.

Allosteric regulations are a natural example of control loops, such as feedback from downstream products or feedforward from upstream substrates. Long-range allostery is especially important in cell signaling. Allosteric regulation is also particularly important in the cell's ability to adjust enzyme activity.

The term allostery comes from the Greek allos (aλλoς), "other," and stereos (στερεος), "solid (object)." This is in reference to the fact that the regulatory site of an allosteric protein is physically distinct from its active site.

Models of Allosteric Regulation

A - Active site

B - Allosteric site

C - Substrate

D - Inhibitor

E - Enzyme

This is a diagram of allosteric regulation of an enzyme.

Most allosteric effects can be explained by the concerted MWC model put forth by Monod, Wyman, and Changeux, or by the sequential model described by Koshland, Nemethy, and Filmer. Both postulate that enzyme subunits exist in one of two conformations, tensed

(T) or relaxed (R), and that relaxed subunits bind substrate more readily than those in the tense state. The two models differ most in their assumptions about subunit interaction and the preexistence of both states.

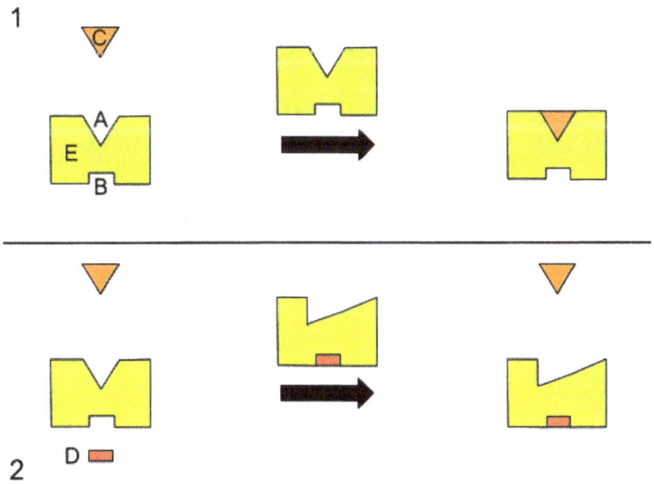

Concerted Model

The concerted model of allostery, also referred to as the symmetry model or MWC model, postulates that enzyme subunits are connected in such a way that a conformational change in one subunit is necessarily conferred to all other subunits. Thus, all subunits must exist in the same conformation. The model further holds that, in the absence of any ligand (substrate or otherwise), the equilibrium favors one of the conformational states, T or R. The equilibrium can be shifted to the R or T state through the binding of one ligand (the allosteric effector or ligand) to a site that is different from the active site (the allosteric site).

Sequential Model

The sequential model of allosteric regulation holds that subunits are not connected in such a way that a conformational change in one induces a similar change in the others. Thus, all enzyme subunits do not necessitate the same conformation. Moreover, the sequential model dictates that molecules of a substrate bind via an induced fit protocol. In general, when a subunit randomly collides with a molecule of substrate, the active site, in essence, forms a glove around its substrate. While such an induced fit converts a subunit from the tensed state to relaxed state, it does not propagate the conformational change to adjacent subunits. Instead, substrate-binding at one subunit only slightly alters the structure of other subunits so that their binding sites are more receptive to substrate. To summarize:

- subunits need not exist in the same conformation

- molecules of substrate bind via induced-fit protocol
- conformational changes are not propagated to all subunits

Morpheein Model

The morpheein model of allosteric regulation is a dissociative concerted model.

A morpheein is a homo-oligomeric structure that can exist as an ensemble of physiologically significant and functionally different alternate quaternary assemblies. Transitions between alternate morpheein assemblies involve oligomer dissociation, conformational change in the dissociated state, and reassembly to a different oligomer. The required oligomer disassembly step differentiates the morpheein model for allosteric regulation from the classic MWC and KNF models. Porphobilinogen synthase (PBGS) is the prototype morpheein.

Ensemble Models

Ensemble models of allosteric regulation enumerate an allosteric system's statistical ensemble as a function of its potential energy function, and then relate specific statistical measurements of allostery to specific energy terms in the energy function (such as an intermolecular salt bridge between two domains). Ensemble models like the Ensemble Allosteric Model and Allosteric Ising Model assume that each domain of the system can adopt two states similar to the MWC model. The allostery landscape model introduced by Cuendet, Weinstein, and LeVine allows for the domains to have any number of states and the contribution of a specific molecular interaction to a given allosteric coupling can be estimated using a rigorous set of rules. Molecular dynamics simulations can be used to estimate a systems's statistical ensemble so that it can be analyzed with the allostery landscape model.

ALLOSTERIC RESOURCES ONLINE

Allosteric Database

Allostery is a direct and efficient means for regulation of biological macromolecule function, produced by the binding of a ligand at an allosteric site topographically distinct from the orthosteric site. Due to the often high receptor selectivity and lower target-based toxicity, allosteric regulation is also expected to play an increasing role in drug discovery and bioengineering. The AlloSteric Database (ASD, http://mdl.shsmu.edu.cn/ASD) provides a central resource for the display, search and analysis of the structure, function and related annotation for allosteric molecules. Currently, ASD contains allosteric proteins from more than 100 species and modulators in three categories (activators, inhibitors, and regulators). Each protein is annotated with detailed description of allostery, biological process and related diseases, and each modulator with binding affinity, physicochemical properties and therapeutic area. Integrating the information of allosteric proteins in ASD should allow the prediction of allostery for unknown proteins, to be followed with experimental validation. In addition, modulators curated in ASD can be used to investigate potential allosteric targets for a query compound, and can help chemists to implement structure modifications for novel allosteric drug design.

Allosteric Residues and Their Prediction using the STRESS Web Server

Not all protein residues play equally important roles in allosteric regulation. The identification of residues that are essential to allostery (so-called "allosteric residues") has been the focus of many studies, especially within the last decade. In part, this growing interest is a result of their general importance in protein science, but also because allosteric residues may be exploited in biomedical contexts. Pharmacologically important proteins with difficult-to-target sites may yield to approaches in which one alternatively targets easier-to-reach residues that are capable of allosterically regulating the primary site of interest. These residues can broadly be classified as surface- and interior-allosteric amino acids. Allosteric sites at the surface generally play regulatory roles that are fundamentally distinct from those within the interior; surface residues may serve as receptors or effector sites in allosteric signal transmission, whereas those within the interior may act to transmit such signals. STRucturally-identified ESSential residues (STRESS, http://stress.molmovdb.org) is a web tool that enables users to submit their own protein structures of interest in order to predict both surface- and interior-allosteric residues in an algorithmically efficient manner. The software behind this server employs 3D structures to build models of conformational change in order to perform predictions.

ALLOSTERIC MODULATION

Positive Modulation

Positive allosteric modulation (also known as allosteric activation) occurs when the binding of one ligand enhances the attraction between substrate molecules and other binding sites. An example is the binding of oxygen molecules to hemoglobin, where oxygen is effectively both the substrate and the effector. The allosteric, or "other", site is the active site of an adjoining protein subunit. The binding of oxygen to one subunit induces a conformational change in that subunit that interacts with the remaining active sites to enhance their oxygen affinity. Another example of allosteric activation is seen in cytosolic IMP-GMP specific 5'-nucleotidase II (cN-II), where the affinity for substrate GMP increases upon GTP binding at the dimer interface

Negative Modulation

Negative allosteric modulation (also known as allosteric inhibition) occurs when the binding of one ligand decreases the affinity for substrate at other active sites. For example, when 2,3-BPG binds to an allosteric site on hemoglobin, the affinity for oxygen of all subunits decreases. This is when a regulator is absent from the binding site.

Direct thrombin inhibitors provides an excellent example of negative allosteric modulation. Allosteric inhibitors of thrombin have been discovered which could potentially be used as anticoagulants.

Another example is strychnine, a convulsant poison, which acts as an allosteric inhibitor of the glycine receptor. Glycine is a major post-synaptic inhibitory neurotransmitter in mammalian spinal cord and brain stem. Strychnine acts at a separate binding site on the glycine receptor in an allosteric manner; i.e., its binding lowers the affinity of the glycine receptor for glycine. Thus, strychnine inhibits the action of an inhibitory transmitter, leading to convulsions.

Another instance in which negative allosteric modulation can be seen is between ATP and the enzyme Phosphofructokinase within the negative feedback loop that regulates glycolysis. Phosphofructokinase (generally referred to as PFK) is an enzyme that catalyses the third step of glycolysis: the phosphorylation of Fructose-6-phosphate into Fructose 1,6-bisphosphate. PFK can be allosterically inhibited by high levels of ATP within the cell. When ATP levels are high, ATP will bind to an allosteoric site on phosphofructokinase, causing a change in the enzyme's three-dimensional shape. This change causes its affinity for substrate (fructose-6-phosphate and ATP) at the active site to decrease, and the enzyme is deemed inactive. This causes glycolysis to cease when ATP levels are high, thus conserving the body's glucose and maintaining balanced levels of cellular ATP. In this way, ATP serves as a negative allosteric modulator for PFK, despite the fact that it is also a substrate of the enzyme.

TYPES

Homotropic

A homotropic allosteric modulator is a substrate for its target enzyme, as well as a regulatory molecule of the enzyme's activity. It is typically an activator of the enzyme. For example, O2 and CO are homotropic allosteric modulators of hemoglobin.

Heterotropic

A heterotropic allosteric modulator is a regulatory molecule that is not the enzyme's substrate. It may be either an activator or an inhibitor of the enzyme. For example, H+, CO2, and 2,3-bisphosphoglycerate are heterotropic allosteric modulators of hemoglobin.

Some allosteric proteins can be regulated by both their substrates and other molecules. Such proteins are capable of both homotropic and heterotropic interactions.

NON-REGULATORY ALLOSTERY

A non-regulatory allosteric site refers to any non-regulatory component of an enzyme (or any protein), that is not itself an amino acid. For instance, many enzymes require sodium binding to ensure proper function. However, the sodium does not necessarily act as a regulatory subunit; the sodium is always present and there are no known biological processes to add/remove sodium to regulate enzyme activity. Non-regulatory allostery could comprise any other ions besides sodium (calcium, magnesium, zinc), as well as other chemicals and possibly vitamins.

PHARMACOLOGY

Allosteric modulation of a receptor results from the binding of allosteric modulators at a different site (a "regulatory site") from that of the endogenous ligand (an "active site") and enhances or inhibits the effects of the endogenous ligand. Under normal circumstances, it acts by causing a conformational change in a receptor molecule, which results in a change in the binding affinity of the ligand. In this way, an allosteric ligand modulates the receptor's activation by its primary (orthosteric) ligand, and can be thought to act like a dimmer switch in an electrical circuit, adjusting the intensity of the response.

For example, the GABAA receptor has two active sites that the neurotransmitter gamma-aminobutyric acid (GABA) binds, but also has benzodiazepine and general anaesthetic agent regulatory binding sites. These regulatory sites can each produce positive allosteric modulation, potentiating the activity of GABA. Diazepam is an agonist at the benzodiazepine regulatory site, and its antidote flumazenil is an antagonist.

More recent examples of drugs that allosterically modulate their targets include the calcium-mimicking cinacalcet and the HIV treatment maraviroc.

Allosteric Sites as Drug Targets

Allosteric sites may represent a novel drug target. There are a number of advantages in using allosteric modulators as preferred therapeutic agents over classic orthosteric ligands. For example, G protein-coupled receptor (GPCR) allosteric binding sites have not faced the same evolutionary pressure as orthosteric sites to accommodate an endogenous ligand, so are more diverse. Therefore, greater GPCR selectivity may be obtained by targeting allosteric sites. This is particularly useful for GPCRs where selective orthosteric therapy has been difficult because of sequence conservation of the orthosteric site across receptor subtypes. Also, these modulators have a decreased potential for toxic effects, since modulators with limited co-operativity will have a ceiling level to their effect, irrespective of the administered dose. Another type of pharmacological selectivity that is unique to allosteric modulators is based on co-operativity. An allosteric modulator may display neutral co-operativity with an orthosteric ligand at all subtypes of a given receptor except the subtype of interest, which is termed "absolute subtype selectivity". If an allosteric modulator does not possess appreciable efficacy, it can provide another powerful therapeutic advantage over orthosteric ligands, namely the ability to selectively tune up or down tissue responses only when the endogenous agonist is present. Oligomer-specific small molecule binding sites are drug targets for medically relevant morpheeins.

COOPERATIVITY

Cooperativity is a phenomenon displayed by systems involving identical or near-identical elements, which act dependently of each other, relative to a hypothetical standard non-interacting system in which the individual elements are acting independently. One manifestation of this is enzymes or receptors that have multiple binding sites where the affinity of the binding sites for a ligand is apparently increased, positive cooperativity, or decreased, negative cooperativity, upon the binding of a ligand to a binding site. For example, when an oxygen atom binds to one of hemoglobin's four binding sites, the affinity to oxygen of the three remaining available binding sites increases; i.e. oxygen is more likely to bind to a hemoglobin bound to one oxygen than to an unbound hemoglobin. This is referred to as cooperative binding. We also see cooperativity in large chain molecules made of many identical (or nearly identical) subunits (such as DNA, proteins, and phospholipids), when such molecules undergo phase transitions such as melting, unfolding or unwinding. This is referred to as subunit cooperativity. However, the definition of cooperativity based on apparent increase or decrease in affinity to successive ligand binding steps is problematic, as the concept of "energy" must always be defined relative to a standard state. When we say that the affinity is increased upon binding of one ligand, it is empirically unclear what we mean since a non-cooperative binding curve is required to rigorously define binding energy and hence also affinity. A much more general and useful definition of positive cooperativity is: A process involving multiple identical incremental steps, in which intermediate states are statistically underrepresented relative to a hypothetical standard system (null hypothesis) where the steps occur independently of each other. Likewise, a definition of negative cooperativity would be a process involving multiple identical incremental steps, in which the intermediate states are overrepresented relative to a hypothetical standard state in which individual steps occur independently. These latter definitions for positive and negative cooperativity easily encompass all processes which we call "cooperative", including conformational transitions in large molecules (such as proteins) and even psychological phenomena of large numbers of people (which can act independently of each other, or in a co-operative fashion).

Cooperative Binding

When a substrate binds to one enzymatic subunit, the rest of the subunits are stimulated and become active. Ligands can either have positive cooperativity, negative cooperativity, or non-cooperativity.

An example of positive cooperativity is the binding of oxygen to hemoglobin. One oxygen molecule can bind to the ferrous iron of a heme molecule in each of the four chains of a hemoglobin molecule. Deoxy-hemoglobin has a relatively low affinity for oxygen, but when one molecule binds to a single heme, the oxygen affinity increases, allowing the second molecule to bind more easily, and the third and fourth even more easily. The oxygen affinity of 3-oxy-hemoglobin is ~300 times greater than that of deoxy-hemoglobin. This behavior leads the affinity curve of hemoglobin to be sigmoidal, rather than hyperbolic as with the monomeric myoglobin. By the same process, the ability for hemoglobin to lose oxygen increases as fewer oxygen molecules are bound. See also Oxygen-hemoglobin dissociation curve.

Negative cooperativity means that the opposite will be true; as ligands bind to the protein, the protein's affinity for the ligand will decrease, i.e. it becomes less likely for the ligand to bind to the protein. An example of this occurring is the relationship between glyceraldehyde-3-phosphate and the enzyme glyceraldehyde-3-phosphate dehydrogenase.

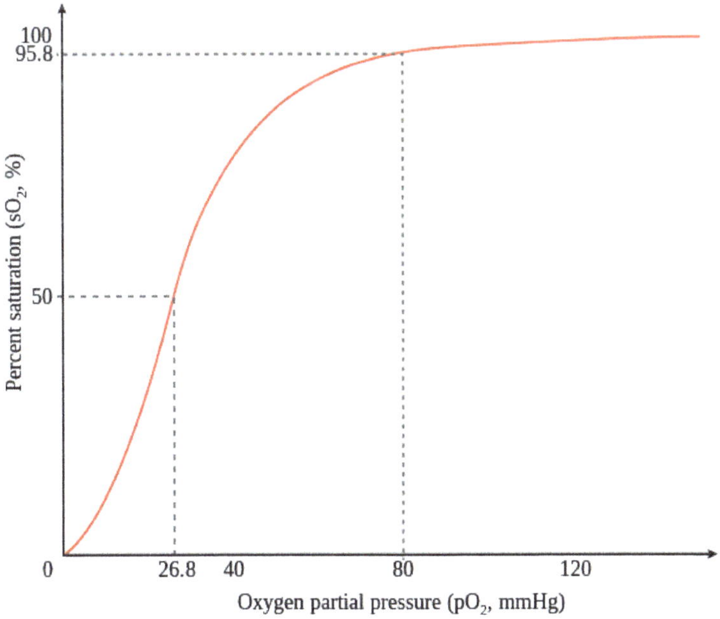

The sigmoidal shape of hemoglobin's oxygen-dissociation curve results from cooperative binding of oxygen to hemoglobin.

Homotropic cooperativity refers to the fact that the molecule causing the cooperativity is the one that will be affected by it. Heterotropic cooperativity is where a third party substance causes the change in affinity.

Subunit Cooperativity

Cooperativity is not only a phenomenon of ligand binding, but also applies anytime energetic interactions make it easier or more difficult for something to happen involving multiple units as opposed to with single units. (That is, easier or more difficult compared with what is expected when only accounting for the addition of multiple units). For example, unwinding of DNA involves cooperativity: Portions of DNA must unwind in order for DNA to carry out replication, transcription and recombination. Positive cooperativity among adjacent DNA nucleotides makes it easier to unwind a whole group of adjacent nucleotides than it is to unwind the same number of nucleotides spread out along the DNA chain. The cooperative unit size is the number of adjacent bases that tend to unwind as a single unit due to the effects of positive cooperativity. This phenomenon applies to other types of chain molecules as well, such as the folding and unfolding of proteins and in the "melting" of phospholipid chains that make up the membranes of cells. Subunit cooperativity is measured on the relative scale known as Hill's Constant.

The Hill Equation

A simple and widely used model for molecular interactions is the Hill Equation. This provides a way to quantify cooperative binding by describing the fraction of saturated ligand binding sites as a function of the ligand concentration.

Entropy and Cooperativity

In all of the above types of cooperativity, entropy plays a role. For example, in the case of oxygen binding to hemoglobin, the first oxygen has four different available binding sites. This represents a state of higher entropy compared to a fourth oxygen having one available binding site. Thus, in transition from the unbound to the bound state, the first oxygen must overcome a larger entropy change than the last oxygen in order to bind to the hemoglobin.

ENZYME INHIBITOR

An enzyme inhibitor is a molecule that binds to an enzyme and decreases its activity. Since blocking an enzyme's activity can kill a pathogen or correct a metabolic imbalance, many drugs are enzyme inhibitors. They are also used in pesticides. Not all molecules that bind to enzymes are inhibitors; enzyme activators bind to enzymes and increase their enzymatic activity, while enzyme substrates bind and are converted to products in the normal catalytic cycle of the enzyme.

The binding of an inhibitor can stop a substrate from entering the enzyme's active site and/or hinder the enzyme from catalyzing its reaction. Inhibitor binding is either reversible or irreversible. Irreversible inhibitors usually react with the enzyme and change it chemically (e.g. via covalent bond formation). These inhibitors modify key amino acid residues needed for enzymatic activity. In contrast, reversible inhibitors bind non-covalently and different types of inhibition are produced depending on whether these inhibitors bind to the enzyme, the enzyme-substrate complex, or both.

Many drug molecules are enzyme inhibitors, so their discovery and improvement is an active area of research in biochemistry and pharmacology. A medicinal enzyme inhibitor is often judged by its specificity (its lack of binding to other proteins) and its potency (its dissociation constant, which indicates the concentration needed to inhibit the enzyme). A high specificity and potency ensure that a drug will have few side effects and thus low toxicity.

Enzyme inhibitors also occur naturally and are involved in the regulation of metabolism. For example, enzymes in a metabolic pathway can be inhibited by downstream products. This type of negative feedback slows the production line when products begin to build up and is an important way to maintain homeostasis in a cell. Other cellular enzyme inhibitors are proteins that specifically bind to and inhibit an enzyme target. This can help control enzymes that may be damaging to a cell, like proteases or nucleases. A well-characterised example of this is the ribonuclease inhibitor, which binds to ribonucleases in one of the tightest known protein–protein interactions. Natural enzyme inhibitors can also be poisons and are used as defences against predators or as ways of killing prey.

Reversible Inhibitors

Types of Reversible Inhibitors

Reversible inhibitors attach to enzymes with non-covalent interactions such as hydrogen bonds, hydrophobic interactions and ionic bonds. Multiple weak bonds between the inhibitor and the active site combine to produce strong and specific binding. In contrast to substrates and irreversible inhibitors, reversible inhibitors generally do not undergo chemical reactions when bound to the enzyme and can be easily removed by dilution or dialysis.

There are four kinds of reversible enzyme inhibitors. They are classified according to the effect of varying the concentration of the enzyme's substrate on the inhibitor.

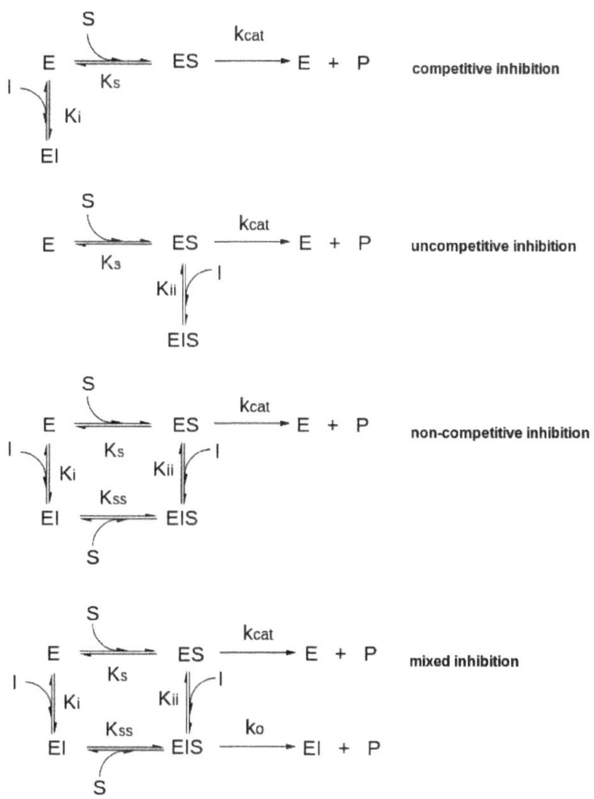

Types of inhibition. This classification was introduced by W.W. Cleland.

- In **competitive inhibition**, the substrate and inhibitor cannot bind to the enzyme at the same time, as shown in the figure on the right. This usually results from the inhibitor having an affinity for the active site of an enzyme where the substrate also binds; the substrate and inhibitor *compete* for access to the enzyme's active site. This type of inhibition can be overcome by sufficiently high concentrations of substrate (V_{max} remains constant), i.e., by out-competing the inhibitor. However, the apparent K_m will increase as it takes a higher concentration of the substrate to reach the K_m point, or half the V_{max}. Competitive inhibitors are often similar in structure to the real substrate (see examples below).
- In **uncompetitive inhibition**, the inhibitor binds only to the substrate-enzyme complex. This type of inhibition causes V_{max} to decrease (maximum velocity decreases as a result of removing activated complex) and K_m to decrease (due to better binding efficiency as a result of Le Chatelier's principle and the effective elimination of the ES complex thus decreasing the K_m which indicates a higher binding affinity).
- In **non-competitive inhibition**, the binding of the inhibitor to the enzyme reduces its activity but does not affect the binding of substrate. As a result, the extent of inhibition depends only on the concentration of the inhibitor. V_{max} will decrease due to the inability for the reaction to proceed as efficiently, but K_m will remain the same as the actual binding of the substrate, by definition, will still function properly.
- In **mixed inhibition**, the inhibitor can bind to the enzyme at the same time as the enzyme's substrate. However, the binding of the inhibitor affects the binding of the substrate, and vice versa. This type of inhibition can be reduced, but not overcome by increasing concentrations of substrate. Although it is possible for mixed-type inhibitors to bind in the active site, this type of inhibition generally results from an allosteric effect where the inhibitor binds to a different site on an enzyme. Inhibitor binding to this allosteric site changes the conformation (i.e., tertiary structure or three-dimensional shape) of the enzyme so that the affinity of the substrate for the active site is reduced.

Quantitative Description of Reversible Inhibition

Reversible inhibition can be described quantitatively in terms of the inhibitor's binding to the enzyme and to the enzyme-substrate complex, and its effects on the kinetic constants of the enzyme. In the classic Michaelis-Menten scheme below, an enzyme (E) binds to its substrate (S) to form the enzyme–substrate complex ES. Upon catalysis, this complex breaks down to release product P and free enzyme. The inhibitor (I) can bind to either E or ES with the dissociation constants Ki or Ki', respectively.

- Competitive inhibitors can bind to E, but not to ES. Competitive inhibition increases K_m (i.e., the inhibitor interferes with substrate binding), but does not affect V_{max} (the inhibitor does not hamper catalysis in ES because it cannot bind to ES).
- Uncompetitive inhibitors bind to ES. Uncompetitive inhibition decreases both K_m' and 'V_{max}. *The inhibitor affects substrate binding by increasing the enzyme's affinity for the substrate (decreasing* K_m*) as well as hampering catalysis (decreases* V_{max}*).*

- Non-competitive inhibitors have identical affinities for E and ES ($K_i = K_i$'). Non-competitive inhibition does not change K_m(i.e., it does not affect substrate binding) but decreases V_{max} (i.e., inhibitor binding hampers catalysis).
- Mixed-type inhibitors bind to both E and ES, but their affinities for these two forms of the enzyme are different ($K_i \neq K_i$'). Thus, mixed-type inhibitors interfere with substrate binding (increase K_m) and hamper catalysis in the ES complex (decrease V_{max}).

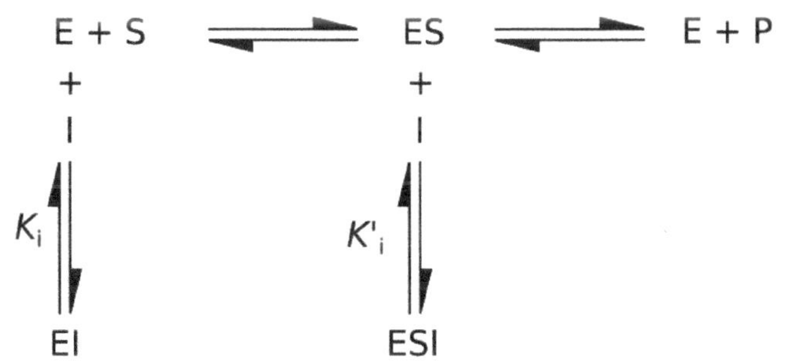

Kinetic scheme for reversible enzyme inhibitors

When an enzyme has multiple substrates, inhibitors can show different types of inhibition depending on which substrate is considered. This results from the active site containing two different binding sites within the active site, one for each substrate. For example, an inhibitor might compete with substrate A for the first binding site, but be a non-competitive inhibitor with respect to substrate B in the second binding site.

Measuring the Dissociation Constants of a Reversible Inhibitor

As noted above, an enzyme inhibitor is characterised by its two dissociation constants, Ki and Ki', to the enzyme and to the enzyme-substrate complex, respectively. The enzyme-inhibitor constant Ki can be measured directly by various methods; one extremely accurate method is isothermal titration calorimetry, in which the inhibitor is titrated into a solution of enzyme and the heat released or absorbed is measured. However, the other dissociation constant Ki' is difficult to measure directly, since the enzyme-substrate complex is short-lived and undergoing a chemical reaction to form the product. Hence, Ki' is usually measured indirectly, by observing the enzyme activity under various substrate and inhibitor concentrations, and fitting the data to a modified Michaelis–Menten equation

$$V = \frac{V_{max}[S]}{\alpha K_m + \alpha'[S]} = \frac{(1/\alpha')V_{max}[S]}{(\alpha/\alpha')K_m + [S]}$$

where the modifying factors α and α' are defined by the inhibitor concentration and its two dissociation constants

$$\alpha = 1 + \frac{[I]}{K_i}$$

$$\alpha' = 1 + \frac{[I]}{K_i'}.$$

Thus, in the presence of the inhibitor, the enzyme's effective Km and Vmax become (α/α')Km and (1/α')Vmax, respectively. However, the modified Michaelis-Menten equation assumes that binding of the inhibitor to the enzyme has reached equilibrium, which may be a very slow process for inhibitors with sub-nanomolar dissociation constants. In these cases, it is usually more practical to treat the tight-binding inhibitor as an irreversible inhibitor (see below); however, it can still be possible to estimate Ki' kinetically if Ki is measured independently.

The effects of different types of reversible enzyme inhibitors on enzymatic activity can be visualized using graphical representations of the Michaelis–Menten equation, such as Lineweaver–Burk and Eadie-Hofstee plots. For example, in the Lineweaver–Burk plots at the right, the competitive inhibition lines intersect on the y-axis, illustrating that such inhibitors do not affect Vmax. Similarly, the non-competitive inhibition lines intersect on the x-axis, showing these inhibitors do not affect Km. However, it can be difficult to estimate Ki and Ki' accurately from such plots, so it is advisable to estimate these constants using more reliable nonlinear regression methods, as described above.

Reversible Inhibitors

Traditionally reversible enzyme inhibitors have been classified as competitive, uncompetitive, or non-competitive, according to their effects on Km and Vmax. These different effects result from the inhibitor binding to the enzyme E, to the enzyme–substrate complex ES, or to both, respectively. The division of these classes arises from a problem in their derivation and results in the need to use two different binding constants for one binding event. The binding of an inhibitor and its effect on the enzymatic activity are two distinctly different things, another problem the traditional equations fail to acknowledge. In noncompetitive inhibition the binding of the inhibitor results in 100% inhibition of the enzyme only, and fails to consider the possibility of anything in between. The common form of the inhibitory term also obscures the relationship between the inhibitor binding to the enzyme and its relationship to any other binding term be it the Michaelis–Menten equation or a dose response curve associated with ligand receptor binding. To demonstrate the relationship the following rearrangement can be made:

$$\frac{V_{\max}}{1 + \frac{[\mathrm{I}]}{K_i}} = \frac{V_{\max}}{\frac{[\mathrm{I}] + K_i}{K_i}}$$

Adding zero to the bottom ([I]-[I])

$$\cfrac{V_{\max}}{\cfrac{[\mathrm{I}]+K_i}{[\mathrm{I}]+K_i-[\mathrm{I}]}}$$

Dividing by [I]+Ki

$$\cfrac{V_{\max}}{\cfrac{1}{1-\cfrac{[\mathrm{I}]}{[\mathrm{I}]+K_i}}}=V_{\max}-V_{\max}\frac{[\mathrm{I}]}{[\mathrm{I}]+K_i}$$

This notation demonstrates that similar to the Michaelis–Menten equation,where the rate of reaction depends on the percent of the enzyme population interacting with substrate.

fraction of the enzyme population bound by substrate

$$\frac{[\mathrm{S}]}{[\mathrm{S}]+K_m}$$

fraction of the enzyme population bound by inhibitor

$$\frac{[\mathrm{I}]}{[\mathrm{I}]+K_i}$$

the effect of the inhibitor is a result of the percent of the enzyme population interacting with inhibitor. The only problem with this equation in its present form is that it assumes absolute inhibition of the enzyme with inhibitor binding, when in fact there can be a wide range of effects anywhere from 100% inhibition of substrate turn over to just >0%. To account for this the equation can be easily modified to allow for different degrees of inhibition by including a delta V_{max} term.

$$V_{\max}-\Delta V_{\max}\frac{[\mathrm{I}]}{[\mathrm{I}]+K_i}$$

or

$$V_{\max 1}-(V_{\max 1}-V_{\max 2})\frac{[\mathrm{I}]}{[\mathrm{I}]+K_i}$$

This term can then define the residual enzymatic activity present when the inhibitor is interacting with individual enzymes in the population. However the inclusion of this term has the added value of allowing for the possibility of activation if the secondary Vmax term turns out to be higher than the initial term. To account for the possibly of activation as well the notation can then be rewritten replacing the inhibitor "I" with a modifier term denoted here as "X".

$$V_{\max 1}-(V_{\max 1}-V_{\max 2})\frac{[\mathrm{X}]}{[\mathrm{X}]+K_x}$$

While this terminology results in a simplified way of dealing with kinetic effects relating to the maximum velocity of the Michaelis–Menten equation, it highlights potential problems with the term used to describe effects relating to the Km. The Km relating to the affinity of the enzyme for the substrate should in most cases relate to potential changes in the binding site of the enzyme which would directly result from enzyme inhibitor interactions. As such a term similar to the one proposed above to modulate Vmaxshould be appropriate in most situations:

$$K_{m1} - (K_{m1} - K_{m2})\frac{[X]}{[X] + K_x}$$

Special Cases

- The mechanism of **partially competitive inhibition** is similar to that of non-competitive, except that the EIS complex has catalytic activity, which may be lower or even higher (partially competitive activation) than that of the enzyme–substrate (ES) complex. This inhibition typically displays a lower V_{max}, but an unaffected K_m value.
- **Uncompetitive inhibition** occurs when the inhibitor binds only to the enzyme–substrate complex, not to the free enzyme; the EIS complex is catalytically inactive. This mode of inhibition is rare and causes a decrease in both V_{max} and the K_m value.
- **Substrate and product inhibition** is where either the substrate or product of an enzyme reaction inhibit the enzyme's activity. This inhibition may follow the competitive, uncompetitive or mixed patterns. In substrate inhibition there is a progressive decrease in activity at high substrate concentrations. This may indicate the existence of two substrate-binding sites in the enzyme. At low substrate, the high-affinity site is occupied and normal kinetics are followed. However, at higher concentrations, the second inhibitory site becomes occupied, inhibiting the enzyme. Product inhibition is often a regulatory feature in metabolism and can be a form of negative feedback.
- **Slow-tight inhibition** occurs when the initial enzyme–inhibitor complex EI undergoes isomerisation to a second more tightly held complex, EI*, but the overall inhibition process is reversible. This manifests itself as slowly increasing enzyme inhibition. Under these conditions, traditional Michaelis–Menten kinetics give a false value for K_i, which is time–dependent. The true value of K_i can be obtained through more complex analysis of the on (k_{on}) and off (k_{off}) rate constants for inhibitor association. See irreversible inhibition below for more information.

Examples of Reversible Inhibitors

As enzymes have evolved to bind their substrates tightly, and most reversible inhibitors bind in the active site of enzymes, it is unsurprising that some of these inhibitors are strikingly similar in structure to the substrates of their targets. An example of these substrate mimics are the protease inhibitors, a very successful class of antiretroviral drugs used to treat HIV. The structure of ritonavir, a protease inhibitor based on a peptide and containing three peptide bonds, is shown on the right. As this drug resembles the protein that is the substrate of the HIV protease, it competes with this substrate in the enzyme's active site.

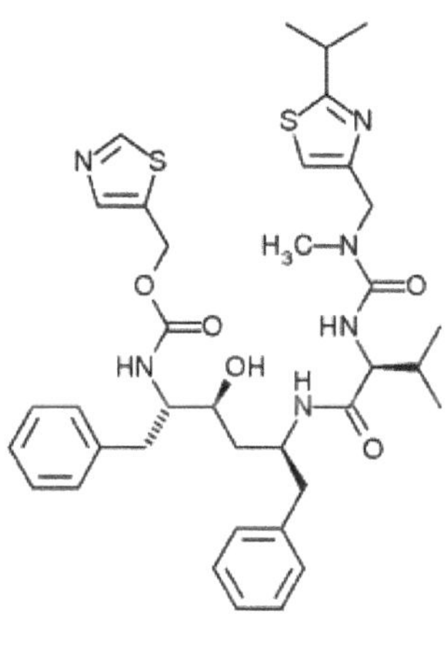

Peptide-based HIV-1 protease inhibitor ritonavir

Enzyme inhibitors are often designed to mimic the transition state or intermediate of an enzyme-catalyzed reaction. This ensures that the inhibitor exploits the transition state stabilising effect of the enzyme, resulting in a better binding affinity (lower Ki) than substrate-based designs. An example of such a transition state inhibitor is the antiviral drug oseltamivir; this drug mimics the planar nature of the ring oxonium ion in the reaction of the viral enzyme neuraminidase.

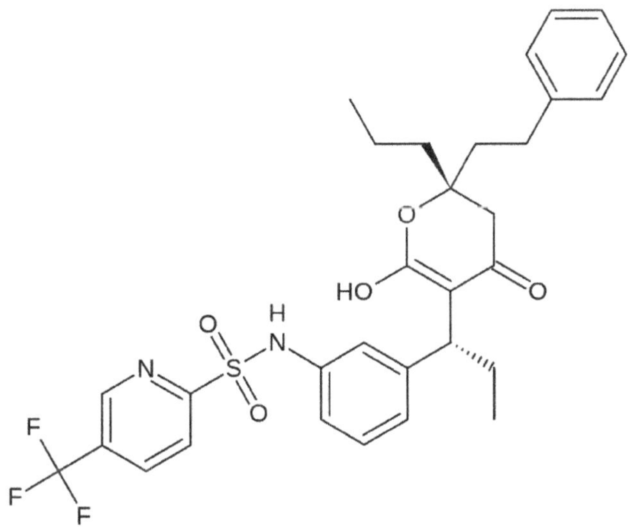

Nonpeptidic HIV-1 protease inhibitor tipranavir.

However, not all inhibitors are based on the structures of substrates. For example, the structure of another HIV protease inhibitor tipranavir is shown on the left. This molecule is not based on a peptide and has no obvious structural similarity to a protein substrate. These non-peptide inhibitors can be more stable than inhibitors containing peptide bonds, because they will not be substrates for peptidases and are less likely to be degraded. In drug design it is important to consider the concentrations of substrates to which the target enzymes are exposed. For example, some protein kinase inhibitors have chemical structures that are similar to adenosine triphosphate, one of the substrates of these enzymes. However, drugs that are simple competitive inhibitors will have to compete with the high concentrations of ATP in the cell. Protein kinases can also be inhibited by competition at the binding sites where the kinases interact with their substrate proteins, and most proteins are present inside cells at concentrations much lower than the concentration of ATP. As a consequence, if two protein kinase inhibitors both bind in the active site with similar affinity, but only one has to compete with ATP, then the competitive inhibitor at the protein-binding site will inhibit the enzyme more effectively.

Irreversible Inhibitors

Types of Irreversible Inhibition

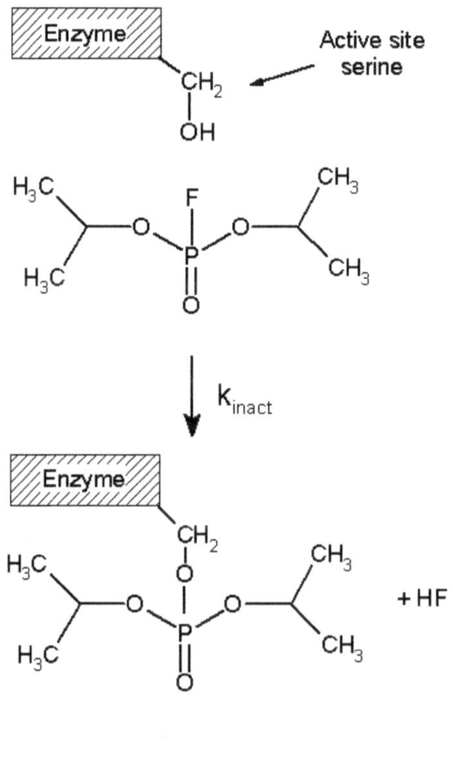

Reaction of the irreversible inhibitor diisopropylfluorophosphate (DFP) with a serine protease.

Irreversible inhibitors usually covalently modify an enzyme, and inhibition can therefore not be reversed. Irreversible inhibitors often contain reactive functional groups such as nitrogen mustards, aldehydes, haloalkanes, alkenes, Michael acceptors, phenyl sulfonates, or fluorophosphonates. These electrophilic groups react with amino acid side chains to form covalent adducts. The residues modified are those with side chains containing nucleophiles such as hydroxyl or sulfhydryl groups; these include the amino acids serine (as in DFP, right), cysteine, threonine, or tyrosine.

Irreversible inhibition is different from irreversible enzyme inactivation. Irreversible inhibitors are generally specific for one class of enzyme and do not inactivate all proteins; they do not function by destroying protein structure but by specifically altering the active site of their target. For example, extremes of pH or temperature usually cause denaturation of all protein structure, but this is a non-specific effect. Similarly, some non-specific chemical treatments destroy protein structure: for example, heating in concentrated hydrochloric acid will hydrolyse the peptide bonds holding proteins together, releasing free amino acids.

Irreversible inhibitors display time-dependent inhibition and their potency therefore cannot be characterised by an IC50 value. This is because the amount of active enzyme at a given concentration of irreversible inhibitor will be different depending on how long the inhibitor is pre-incubated with the enzyme. Instead, kobs/[I] values are used, wherekobs is the observed pseudo-first order rate of inactivation (obtained by plotting the log of % activity vs. time) and [I] is the concentration of inhibitor. The kobs/[I] parameter is valid as long as the inhibitor does not saturate binding with the enzyme (in which case kobs = kinact).

Analysis of Irreversible Inhibition

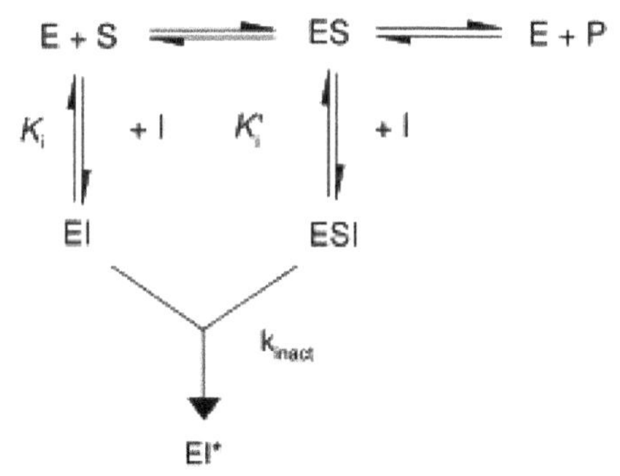

Kinetic scheme for irreversible inhibitors.

As shown in the figure to the left, irreversible inhibitors form a reversible non-covalent complex with the enzyme (EI or ESI) and this then reacts to produce the covalently modified "dead-end complex" EI*. The rate at which EI* is formed is called the inactivation rate or kinact. Since formation of EI may compete with ES, binding of irreversible inhibitors can be prevented by competition either with substrate or with a second, reversible inhibitor. This protection effect is good evidence of a specific reaction of the irreversible inhibitor with the active site.

The binding and inactivation steps of this reaction are investigated by incubating the enzyme with inhibitor and assaying the amount of activity remaining over time. The activity will be decreased in a time-dependent manner, usually following exponential decay. Fitting these data to a rate equation gives the rate of inactivation at this concentration of inhibitor. This is done at several different concentrations of inhibitor. If a reversible EI complex is involved the inactivation rate will be saturable and fitting this curve will give kinact and Ki.

Another method that is widely used in these analyses is mass spectrometry. Here, accurate measurement of the mass of the unmodified native enzyme and the inactivated enzyme gives the increase in mass caused by reaction with the inhibitor and shows the stoichiometry of the reaction. This is usually done using a MALDI-TOF mass spectrometer. In a complementary technique, peptide mass fingerprinting involves digestion of the native and modified protein with a protease such as trypsin. This will produce a set of peptides that can be analysed using a mass spectrometer. The peptide that changes in mass after reaction with the inhibitor will be the one that contains the site of modification.

Special Cases

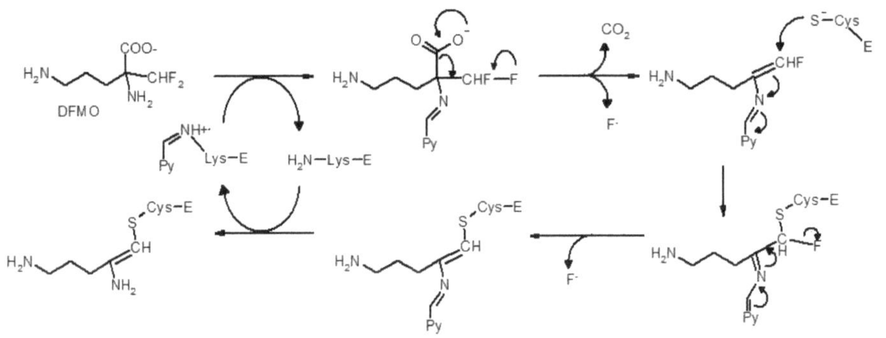

Chemical mechanism for irreversible inhibition of ornithine decarboxylase by DFMO. Pyridoxal 5'-phosphate (Py) and enzyme (E) are not shown. Adapted from

Not all irreversible inhibitors form covalent adducts with their enzyme targets. Some reversible inhibitors bind so tightly to their target enzyme that they are essentially irreversible. These tight-binding inhibitors may show kinetics similar to covalent irreversible inhibitors. In these cases, some of these inhibitors rapidly bind to the enzyme in a low-affinity EI complex and this then undergoes a slower rearrangement to a very tightly bound EI* complex (see figure above). This kinetic behaviour is called slow-binding.This slow rearrangement after binding often involves a conformational change as the enzyme "clamps down" around the inhibitor molecule. Examples of slow-binding inhibitors include some important drugs, such methotrexate, allopurinol, and the activated form of acyclovir.

Examples of Irreversible Inhibitors

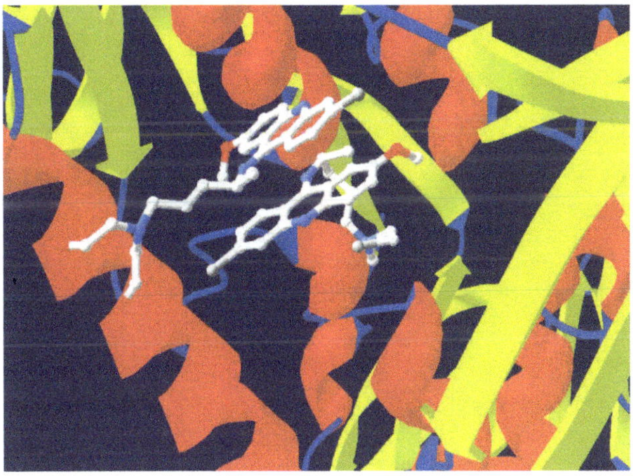

Trypanothione reductase with the lower molecule of an inhibitor bound irreversibly and the upper one reversibly. Created from PDB 1GXF.

Diisopropylfluorophosphate (DFP) is shown as an example of an irreversible protease inhibitor in the figure above right. The enzyme hydrolyses the phosphorus–fluorine bond, but the phosphate residue remains bound to the serine in the active site, deactivating it.Similarly, DFP also reacts with the active site of acetylcholine esterase in the synapses of neurons, and consequently is a potent neurotoxin, with a lethal dose of less than 100 mg.

Suicide inhibition is an unusual type of irreversible inhibition where the enzyme converts the inhibitor into a reactive form in its active site. An example is the inhibitor of polyamine biosynthesis, α-difluoromethylornithine or DFMO, which is an analogue of the amino acid ornithine, and is used to treat African trypanosomiasis (sleeping sickness). Ornithine decarboxylase can catalyse the decarboxylation of DFMO instead of ornithine, as shown above. However, this decarboxylation reaction is followed by the elimination of a fluorine atom, which converts this catalytic intermediate into a conjugated imine, a highly electrophilic species. This reactive form of DFMO then reacts with either a cysteine or lysine residue in the active site to irreversibly inactivate the enzyme.

Since irreversible inhibition often involves the initial formation of a non-covalent EI complex, it is sometimes possible for an inhibitor to bind to an enzyme in more than one way. For example, in the figure showing trypanothione reductase from the human protozoan parasite Trypanosoma cruzi, two molecules of an inhibitor called quinacrine mustard are bound in its active site. The top molecule is bound reversibly, but the lower one is bound covalently as it has reacted with an amino acid residue through its nitrogen mustard group.

Discovery and Design Of Inhibitors

Robots used for the high-throughput screening of chemical libraries to discover new enzyme inhibitors.

New drugs are the products of a long drug development process, the first step of which is often the discovery of a new enzyme inhibitor. In the past the only way to discover these new inhibitors was by trial and error: screening huge libraries of compounds against a target enzyme and hoping that some useful leads would emerge. This brute force approach is still successful and has even been extended by combinatorial chemistry approaches that quickly produce large numbers of novel compounds and high-throughput screening technology to rapidly screen these huge chemical libraries for useful inhibitors.

More recently, an alternative approach has been applied: rational drug design uses the three-dimensional structure of an enzyme's active site to predict which molecules might be inhibitors. These predictions are then tested and one of these tested compounds may be a novel inhibitor. This new inhibitor is then used to try to obtain a structure of the enzyme in an inhibitor/enzyme complex to show how the molecule is binding to the active site, allowing changes to be made to the inhibitor to try to optimise binding. This test and improve cycle is then repeated until a sufficiently potent inhibitor is produced. Computer-based methods of predicting the affinity of an inhibitor for an enzyme are also being developed, such as molecular docking and molecular mechanics.

Uses of Inhibitors

Enzyme inhibitors are found in nature and are also designed and produced as part of pharmacology and biochemistry. Natural poisons are often enzyme inhibitors that have evolved to defend a plant or animal against predators. These natural toxins include some of the most poisonous compounds known. Artificial inhibitors are often used as drugs, but can also be insecticides such as malathion, herbicides such as glyphosate, or disinfectants such as triclosan. Other artificial enzyme inhibitors block acetylcholinesterase, an enzyme which breaks down acetylcholine, and are used as nerve agents in chemical warfare.

Chemotherapy

The most common uses for enzyme inhibitors are as drugs to treat disease. Many of these inhibitors target a human enzyme and aim to correct a pathological condition. However, not all drugs are enzyme inhibitors. Some, such as anti-epileptic drugs, alter enzyme activity by causing more or less of the enzyme to be produced. These effects are called enzyme induction and inhibition and are alterations in gene expression, which is unrelated to the type of enzyme inhibition discussed here. Other drugs interact with cellular targets that are not enzymes, such as ion channels or membrane receptors.

An example of a medicinal enzyme inhibitor is sildenafil (Viagra), a common treatment for male erectile dysfunction. This compound is a potent inhibitor of cGMP specific phosphodiesterase type 5, the enzyme that degrades the signalling molecule cyclic guanosine monophosphate.This signalling molecule triggers smooth muscle relaxation and allows blood flow into the corpus cavernosum, which causes an erection. Since the drug decreases the activity of the enzyme that halts the signal, it makes this signal last for a longer period of time.

Another example of the structural similarity of some inhibitors to the substrates of the enzymes they target is seen in the figure comparing the drug methotrexate to folic acid. Folic acid is a substrate of dihydrofolate reductase, an enzyme involved in making nucleotides that is potently inhibited by methotrexate. Methotrexate blocks the action of dihydrofolate reductase and thereby halts the production of nucleotides. This block of nucleotide biosynthesis is more toxic to rapidly growing cells than non-dividing cells, since a rapidly growing cell has to carry out DNA replication, therefore methotrexate is often used in cancer chemotherapy.

Antibiotics

Drugs also are used to inhibit enzymes needed for the survival of pathogens. For example, bacteria are surrounded by a thick cell wall made of a net-like polymer called peptidoglycan. Many antibiotics such as penicillin and vancomycin inhibit the enzymes that produce and then cross-link the strands of this polymer together. This causes the cell wall to lose strength and the bacteria to burst. In the figure, a molecule of penicillin (shown in a ball-and-stick form) is shown bound to its target, the transpeptidase from the bacteria Streptomyces R61 (the protein is shown as a ribbon-diagram).

Antibiotic drug design is facilitated when an enzyme that is essential to the pathogen's survival is absent or very different in humans. In the example above, humans do not make peptidoglycan, therefore inhibitors of this process are selectively toxic to bacteria. Selective toxicity is also produced in antibiotics by exploiting differences in the structure of the ribosomes in bacteria, or how they make fatty acids.

Metabolic Control

Enzyme inhibitors are also important in metabolic control. Many metabolic pathways in the cell are inhibited by metabolites that control enzyme activity through allosteric regulation or substrate inhibition. A good example is the allosteric regulation of the glycolytic pathway.

This catabolic pathway consumes glucose and produces ATP, NADH and pyruvate. A key step for the regulation of glycolysis is an early reaction in the pathway catalysed by phosphofructokinase-1 (PFK1). When ATP levels rise, ATP binds an allosteric site in PFK1 to decrease the rate of the enzyme reaction; glycolysis is inhibited and ATP production falls. This negative feedback control helps maintain a steady concentration of ATP in the cell. However, metabolic pathways are not just regulated through inhibition since enzyme activation is equally important. With respect to PFK1, fructose 2,6-bisphosphate and ADP are examples of metabolites that are allosteric activators.

Physiological enzyme inhibition can also be produced by specific protein inhibitors. This mechanism occurs in the pancreas, which synthesises many digestive precursor enzymes known as zymogens. Many of these are activated by the trypsin protease, so it is important to inhibit the activity of trypsin in the pancreas to prevent the organ from digesting itself. One way in which the activity of trypsin is controlled is the production of a specific and potent trypsin inhibitor protein in the pancreas. This inhibitor binds tightly to trypsin, preventing the trypsin activity that would otherwise be detrimental to the organ.Although the trypsin inhibitor is a protein, it avoids being hydrolysed as a substrate by the protease by excluding water from trypsin's active site and destabilising the transition state. Other examples of physiological enzyme inhibitor proteins include the barstar inhibitor of the bacterial ribonuclease barnase and the inhibitors of protein phosphatases.

Pesticides

Many pesticides are enzyme inhibitors. Acetylcholinesterase (AChE) is an enzyme found in animals from insects to humans. It is essential to nerve cell function through its mechanism of breaking down the neurotransmitter acetylcholine into its constituents, acetate and choline. This is somewhat unique among neurotransmitters as most, including serotonin, dopamine, and norepinephrine, are absorbed from the synaptic cleft rather than cleaved. A large number of AChE inhibitors are used in both medicine and agriculture. Reversible competitive inhibitors, such as edrophonium, physostigmine, and neostigmine, are used in the treatment of myasthenia gravis and in anaesthesia. The carbamate pesticides are also examples of reversible AChE inhibitors. The organophosphate pesticides such as malathion, parathion, and chlorpyrifos irreversibly inhibit acetylcholinesterase.

The herbicide glyphosate is an inhibitor of 3-phosphoshikimate 1-carboxyvinyltransferase, other herbicides, such as the sulfonylureas inhibit the enzyme acetolactate synthase. Both these enzymes are needed for plants to make branched-chain amino acids. Many other enzymes are inhibited by herbicides, including enzymes needed for the biosynthesis of lipids and carotenoids and the processes of photosynthesis and oxidative phosphorylation.

To discourage seed predators, pulses contain trypsin inhibitors that interfere with digestion.

Natural Poisons

Animals and plants have evolved to synthesise a vast array of poisonous products including secondary metabolites, peptides and proteins that can act as inhibitors. Natural toxins are usually small organic molecules and are so diverse that there are probably natural inhibitors for most metabolic processes. The metabolic processes targeted by natural poisons encompass more than enzymes in metabolic pathways and can also include the inhibition of receptor, channel and structural protein functions in a cell. For example, paclitaxel (taxol), an organic molecule found in the Pacific yew tree, binds tightly to tubulin dimers and inhibits their assembly into microtubules in the cytoskeleton.

Many natural poisons act as neurotoxins that can cause paralysis leading to death and have functions for defence against predators or in hunting and capturing prey. Some of these natural inhibitors, despite their toxic attributes, are valuable for therapeutic uses at lower doses. An example of a neurotoxin are the glycoalkaloids, from the plant species in the Solanaceae family (includes potato, tomato and eggplant), that are acetylcholinesterase inhibitors. Inhibition of this enzyme causes an uncontrolled increase in the acetylcholine neurotransmitter, muscular paralysis and then death. Neurotoxicity can also result from the inhibition of receptors; for example, atropine from deadly nightshade (Atropa belladonna) that functions as a competitive antagonist of the muscarinic acetylcholine receptors.

Although many natural toxins are secondary metabolites, these poisons also include peptides and proteins. An example of a toxic peptide is alpha-amanitin, which is found in relatives of the death cap mushroom. This is a potent enzyme inhibitor, in this case preventing the RNA polymerase II enzyme from transcribing DNA. The algal toxin microcystin is also a peptide and is an inhibitor of protein phosphatases. This toxin can contaminate water supplies after algal blooms and is a known carcinogen that can also cause acute liver hemorrhage and death at higher doses.

Proteins can also be natural poisons or antinutrients, such as the trypsin inhibitors (discussed above) that are found in some legumes, as shown in the figure above. A less common class of toxins are toxic enzymes: these act as irreversible inhibitors of their target enzymes and work by chemically modifying their substrate enzymes. An example is ricin, an extremely potent protein toxin found in castor oil beans. This enzyme is a glycosidase that inactivates ribosomes. Since ricin is a catalytic irreversible inhibitor, this allows just a single molecule of ricin to kill a cell.

REFERENCES

1. Shapiro, R; Vallee, BL (1991). "Interaction of human placental ribonuclease with placental ribonuclease inhibitor". Biochemistry. 30 *(8): 2246–55.* doi:10.1021/bi00222a030. PMID 1998683.
2. Berg J., Tymoczko J. and Stryer L. (2002) Biochemistry. W. H. Freeman and Company, ISBN 0-7167-4955-6.
3. Cleland WW (1963). "The kinetics of enzyme-catalyzed reactions with two or more substrates or products. II. Inhibition: nomenclature and theory". Biochim. Biophys. Acta. 67: 173–87. doi:10.1016/0926-6569(63)90226-8. PMID 14021668.
4. Irwin H. Segel, Enzyme Kinetics : Behavior and Analysis of Rapid Equilibrium and Steady-State Enzyme Systems. Wiley–Interscience; New edition (1993), ISBN 0-471-30309-7
5. Holdgate, GA (2001). "Making cool drugs hot: isothermal titration calorimetry as a tool to study binding energetics". BioTechniques. 31 *(1): 164–6, 168, 170 passim. PMID 11464510.*
6. Leatherbarrow, RJ (1990). "Using linear and non-linear regression to fit biochemical data". Trends in Biochemical Sciences. 15 *(12): 455–8.* doi:10.1016/0968-0004(90)90295-M. PMID 2077683.
7. Tseng, SJ; Hsu, JP (1990). "A comparison of the parameter estimating procedures for the Michaelis-Menten model". Journal of Theoretical Biology. 145 *(4): 457–64.* doi:10.1016/S0022-5193(05)80481-3. PMID 2246896.
8. Walsh, R.; Martin, E.; Darvesh, S. (2011). "Limitations of conventional inhibitor classifications". Integrative Biology. Royal Society of Chemistry. 3 *(12): 1197–1201.* doi:10.1039/c1ib00053e. PMID 22038120.
9. Walsh, R.; Martin, E.; Darvesh, S. (2007). "A versatile equation to describe reversible enzyme inhibition and activation kinetics: Modeling β-galactosidase and butyrylcholinesterase". Biochimica et Biophysica Acta (BBA) - General Subjects. 1770 *(5): 733–746. doi:10.1016/j.bbagen.2007.01.001. PMID 17307293.*
10. Walsh, Ryan (2012). "Ch. 17. Alternative Perspectives of Enzyme Kinetic Modeling". In Ekinci, Deniz. Medicinal Chemistry and Drug Design (PDF). InTech. pp. 357–371. ISBN 978-953-51-0513-8.
11. Segel, Irwin H. (1993) Enzyme Kinetics : Behavior and Analysis of Rapid Equilibrium and Steady-State Enzyme Systems. Wiley-Interscience; New edition , ISBN 0-471-30309-7.
12. Dixon, M. Webb, E.C., Thorne, C.J.R. and Tipton K.F., Enzymes (3rd edition) Longman, London (1979) p. 126
13. Hsu, JT; Wang, HC; Chen, GW; Shih, SR (2006). "Antiviral drug discovery targeting to viral proteases". Current pharmaceutical design. 12 *(11): 1301–14.* doi:10.2174/138161206776361110. PMID 16611117.

14. Lew W, Chen X, Kim CU; Chen; Kim (2000). "Discovery and development of GS 4104 (oseltamivir): an orally active influenza neuraminidase inhibitor". Curr. Med. Chem. 7 *(6): 663–72.* doi:10.2174/0929867003374886. PMID 10702632.

15. Fischer PM (2003). "The design, synthesis and application of stereochemical and directional peptide isomers: a critical review". Curr. Protein Pept. Sci. 4 *(5): 339–56.* doi:10.2174/1389203033487054. PMID 14529528.

16. Bogoyevitch, MA; Barr, RK; Ketterman, AJ (2005). "Peptide inhibitors of protein kinases-discovery, characterisation and use". Biochimica et Biophysica Acta. 1754 *(1–2): 79–99. doi:10.1016/j.bbapap.2005.07.025. PMID 16182621.*

17. Lundblad R. L. Chemical Reagents for Protein Modification CRC Press Inc (2004) ISBN 0-8493-1983-8

18. Price, N.; Hames, B. and Rickwood, D. (eds.) (1996) Proteins LabFax Academic Press, ISBN 0-12-564710-7.

19. Adam, GC; Cravatt, BF; Sorensen, EJ (2001). "Profiling the specific reactivity of the proteome with non-directed activity-based probes". Chemistry & Biology. 8 *(1): 81–95.* doi:10.1016/S1074-5521(00)90060-7. PMID 11182321.

20. Maurer, T; Fung, HL (2000). "Comparison of methods for analyzing kinetic data from mechanism-based enzyme inactivation: application to nitric oxide synthase". AAPS PharmSci. 2 *(1): 68–77.* doi:10.1208/ps020108. PMC 2751003. PMID 11741224.

21. Loo JA, DeJohn DE, Du P, Stevenson TI, Ogorzalek Loo RR; Dejohn; Du; Stevenson; Ogorzalek Loo (1999). "Application of mass spectrometry for target identification and characterization". Med Res Rev. 19 *(4): 307–19.* doi:10.1002/(SICI)1098-1128(199907)19:4<307::AID-MED4>3.0.CO;2-2. PMID 10398927.

22. Poulin, R; Lu, L; Ackermann, B; Bey, P; Pegg, AE (1992). "Mechanism of the irreversible inactivation of mouse ornithine decarboxylase by alpha-difluoromethylornithine. Characterization of sequences at the inhibitor and coenzyme binding sites". The Journal of Biological Chemistry. 267 *(1): 150–8. PMID 1730582.*

23. Szedlacsek, SE; Duggleby, RG (1995). "Kinetics of slow and tight-binding inhibitors". Enzyme Kinetics and Mechanism Part D: Developments in Enzyme Dynamics. Methods in Enzymology. 249. pp. 144–80. doi:10.1016/0076-6879(95)49034-5. ISBN 978-0-12-182150-0. PMID 7791610.

24. Stone, SR; Morrison, JF (1986). "Mechanism of inhibition of dihydrofolate reductases from bacterial and vertebrate sources by various classes of folate analogues". Biochimica et Biophysica Acta. 869 *(3): 275–85.* doi:10.1016/0167-4838(86)90067-1. PMID 3511964.

25. Pick, FM; McGartoll, MA; Bray, RC (1971). "Reaction of formaldehyde and of methanol with xanthine oxidase". European Journal of Biochemistry / FEBS. 18 *(1): 65–72. doi:10.1111/j.1432-1033.1971.tb01215.x. PMID 4322209.*

26. Reardon, JE (1989). "Herpes simplex virus type 1 and human DNA polymerase interactions with 2'-deoxyguanosine 5'-triphosphate analogues. Kinetics of incorporation into DNA and induction of inhibition". The Journal of Biological Chemistry. 264 *(32): 19039–44. PMID 2553730.*

27. Cohen, J.A.; Oosterbaan, R.A.; Berends, F. (1967). " Organophosphorus compounds".

Enzyme Structure. Methods in Enzymology. 11. p. 686. doi:10.1016/S0076-6879(67)11085-9. ISBN 978-0-12-181860-9.

28. Brenner, G. M. (2000): Pharmacology. Philadelphia, PA: W.B. Saunders Company. ISBN 0-7216-7757-6

29. Saravanamuthu, A; Vickers, TJ; Bond, CS; Peterson, MR; Hunter, WN; Fairlamb, AH (2004). "Two interacting binding sites for quinacrine derivatives in the active site of trypanothione reductase: a template for drug design". The Journal of Biological Chemistry. 279 *(28): 29493–500. doi:10.1074/jbc.M403187200. PMC 3491871. PMID 15102853.*

30. Koppitz M, Eis K; Eis (2006). "Automated medicinal chemistry". Drug Discov. Today. 11 *(11–12): 561–8. doi:10.1016/j.drudis.2006.04.005. PMID 16713909.*

31. Scapin G (2006). "Structural biology and drug discovery". Curr. Pharm. Des. 12 *(17): 2087–97.* doi:10.2174/138161206777585201. PMID 16796557.

32. Gohlke H, Klebe G; Klebe (August 2002). "Approaches to the description and prediction of the binding affinity of small-molecule ligands to macromolecular receptors". Angew. Chem. Int. Ed. Engl. 41 *(15): 2644–76.* doi:10.1002/1521-3773(20020802)41:15<2644::AID-ANIE2644>3.0.CO;2-O. PMID 12203463.

33. Glen RC, Allen SC; Allen (May 2003). "Ligand-protein docking: cancer research at the interface between biology and chemistry". Curr. Med. Chem. 10 *(9): 763–7.* doi:10.2174/0929867033457809. PMID 12678780.

34. Maggi, M; Filippi, S; Ledda, F; Magini, A; Forti, G (2000). "Erectile dysfunction: from biochemical pharmacology to advances in medical therapy". European Journal of Endocrinology. Bioscientifica. 143 *(2): 143–54. doi:10.1530/eje.0.1430143. PMID 10913932.*

35. McGuire, JJ (2003). "Anticancer antifolates: current status and future directions". Current pharmaceutical design. 9 *(31): 2593–613.* doi:10.2174/1381612033453712. PMID 14529544.

36. Katz, AH; Caufield, CE (2003). "Structure-based design approaches to cell wall biosynthesis inhibitors". Current pharmaceutical design. 9 *(11): 857–66.* doi:10.2174/1381612033455305. PMID 12678870.

37. Okar, DA; Lange, AJ (1999). "Fructose-2,6-bisphosphate and control of carbohydrate metabolism in eukaryotes". BioFactors (Oxford, England). 10 *(1): 1–14. doi:10.1002/biof.5520100101. PMID 10475585.*

38. Price, Nicholas and Stevens, Lewis (1999) Fundamentals of Enzymology, Oxford University Press, ISBN 0-19-850229-X.

39. Smyth, TP (2004). "Substrate variants versus transition state analogues as noncovalent reversible enzyme inhibitors". Bioorganic & Medicinal Chemistry. 12 *(15): 4081–8. doi:10.1016/j.bmc.2004.05.041. PMID 15246086.*

40. Hartley, RW (1989). "Barnase and barstar: two small proteins to fold and fit together". Trends in Biochemical Sciences. 14 *(11): 450–4.* doi:10.1016/0968-0004(89)90104-7. PMID 2696173.

41. Oliver, CJ; Shenolikar, S (1998). "Physiologic importance of protein phosphatase inhibitors". Frontiers in Bioscience. 3: D961–72. PMID 9727084.

42. Tan S, Evans R, Singh B; Evans; Singh (March 2006). "Herbicidal inhibitors of amino acid biosynthesis and herbicide-tolerant crops". Amino Acids. 30 *(2): 195–204.* doi:10.1007/s00726-005-0254-1. PMID 16547651.

43. Duke SO (1990). "Overview of herbicide mechanisms of action". Environ. Health Perspect. Brogan &#38. 87: 263–71. doi:10.2307/3431034. JSTOR 3431034. PMC 1567841. PMID 1980104.

44. Tan, G; Gyllenhaal, C; Soejarto, DD (2006). "Biodiversity as a source of anticancer drugs". Current drug targets. 7 *(3): 265–77.* doi:10.2174/138945006776054942. PMID 16515527.

45. Abal, M; Andreu, JM; Barasoain, I (2003). "Taxanes: microtubule and centrosome targets, and cell cycle dependent mechanisms of action". Current cancer drug targets. 3 *(3): 193–203.* doi:10.2174/1568009033481967. PMID 12769688.

46. Hostettmann, K.; Borloz, A.; Urbain, A.; Marston, A. (2006). "Natural Product Inhibitors of Acetylcholinesterase". Current Organic Chemistry. 10 *(8): 825–847.* doi:10.2174/138527206776894410.

47. Defrates, LJ; Hoehns, JD; Sakornbut, EL; Glascock, DG; Tew, AR (2005). "Antimuscarinic intoxication resulting from the ingestion of moonflower seeds". The Annals of pharmacotherapy. 39 *(1): 173–6. doi:10.1345/aph.1D536. PMID 15572604.*

48. Vetter, J (1998). "Toxins of Amanita phalloides". Toxicon. 36 *(1): 13–24.* doi:10.1016/S0041-0101(97)00074-3. PMID 9604278.

49. Holmes, CF; Maynes, JT; Perreault, KR; Dawson, JF; James, MN (2002). "Molecular enzymology underlying regulation of protein phosphatase-1 by natural toxins". Current medicinal chemistry. 9 *(22): 1981–9.* doi:10.2174/0929867023368827. PMID 12369866.

50. Bischoff, K (2001). "The toxicology of microcystin-LR: occurrence, toxicokinetics, toxicodynamics, diagnosis and treatment". Veterinary and human toxicology. 43 *(5): 294–7. PMID 11577938.*

51. Hartley, MR; Lord, JM (2004). "Cytotoxic ribosome-inactivating lectins from plants". Biochimica et Biophysica Acta. 1701 *(1–2): 1–14. doi:10.1016/j.bbapap.2004.06.004. PMID 15450171.*

CHAPTER 7

Oxidoreductase

Chapter Contents

- **Introduction**
- **Reactions**
- **Nomenclature**
- **Classification**
- **Transferase**
- **Hydrolase**
- **Lyase**
- **Isomerase**
- **Ligase**

Oxidoreductases are a class of enzymes that catalyze oxidoreduction reactions . Oxidoreductases catalyze the transfer of electrons from one molecule (the oxidant) to another molecule (the reductant). Oxidoreductases catalyze reactions similar to the following, A– + B → A + B– where A is the oxidant and B is the reductant. Oxidorecuctases can be oxidases or dehydrogenases. Oxidases are enzymes involved when molecular oxygen acts as an acceptor of hydrogen or electrons. Whereas, dehydrogenases are enzymes that oxidize a substrate by transferring hydrogen to an acceptor that is either NAD+/NADP+ or a flavin enzyme. Other oxidoreductases include peroxidases, hydroxylases, oxygenases, and reductases. Peroxidases are localized in peroxisomes, and catalyzes the reduction of hydrogen peroxide. Hydroxylases add hydroxyl groups to its substrates. Oxygenases incorporate oxygen from molecular oxygen into organic substrates. Reductases catalyze reductions, in most cases reductases can act like an oxidases .

INTRODUCTION

In biochemistry, an oxidoreductase is an enzyme that catalyzes the transfer of electrons from one molecule, the reductant, also called the electron donor, to another, the oxidant, also called the electron acceptor. This group of enzymes usually utilizes NADP or NAD+ as cofactors.

REACTIONS

For example, an enzyme that catalyzed this reaction would be an oxidoreductase:

A– + B → A + B–

In this example, A is the reductant (electron donor) and B is the oxidant (electron acceptor).

In biochemical reactions, the redox reactions are sometimes more difficult to see, such as this reaction from glycolysis:

Pi + glyceraldehyde-3-phosphate + NAD+ → NADH + H+ + 1,3-bisphosphoglycerate

In this reaction, NAD+ is the oxidant (electron acceptor), and glyceraldehyde-3-phosphate is the reductant (electron donor).

NOMENCLATURE

Proper names of oxidoreductases are formed as "donor:acceptor oxidoreductase"; however, other names are much more common. The common name is "donor dehydrogenase" when possible, such as glyceraldehyde-3-phosphate dehydrogenase for the second reaction above. Common names are also sometimes formed as "acceptor reductase", such as NAD+ reductase. "Donor oxidase" is a special case where O2 is the acceptor.

CLASSIFICATION

Oxidoreductases are classified as EC 1 in the EC number classification of enzymes. Oxidoreductases can be further classified into 22 subclasses:

- EC 1.1 includes oxidoreductases that act on the CH-OH group of donors (alcohol oxidoreductases)
- EC 1.2 includes oxidoreductases that act on the aldehyde or oxo group of donors
- EC 1.3 includes oxidoreductases that act on the CH-CH group of donors (CH-CH oxidoreductases)
- EC 1.4 includes oxidoreductases that act on the CH-NH_2 group of donors (Amino acid oxidoreductases, Monoamine oxidase)
- EC 1.5 includes oxidoreductases that act on CH-NH group of donors
- EC 1.6 includes oxidoreductases that act on NADH or NADPH
- EC 1.7 includes oxidoreductases that act on other nitrogenous compounds as donors
- EC 1.8 includes oxidoreductases that act on a sulfur group of donors
- EC 1.9 includes oxidoreductases that act on a heme group of donors
- EC 1.10 includes oxidoreductases that act on diphenols and related substances

as donors

- EC 1.11 includes oxidoreductases that act on peroxide as an acceptor (peroxidases)
- EC 1.12 includes oxidoreductases that act on hydrogen as donors
- EC 1.13 includes oxidoreductases that act on single donors with incorporation of molecular oxygen (oxygenases)
- EC 1.14 includes oxidoreductases that act on paired donors with incorporation of molecular oxygen
- EC 1.15 includes oxidoreductases that act on superoxide radicals as acceptors
- EC 1.16 includes oxidoreductases that oxidize metal ions
- EC 1.17 includes oxidoreductases that act on CH or CH_2 groups
- EC 1.18 includes oxidoreductases that act on iron-sulfur proteins as donors
- EC 1.19 includes oxidoreductases that act on reduced flavodoxin as a donor
- EC 1.20 includes oxidoreductases that act on phosphorus or arsenic in donors
- EC 1.21 includes oxidoreductases that act on X-H and Y-H to form an X-Y bond
- EC 1.97 includes other oxidoreductases

TRANSFERASE

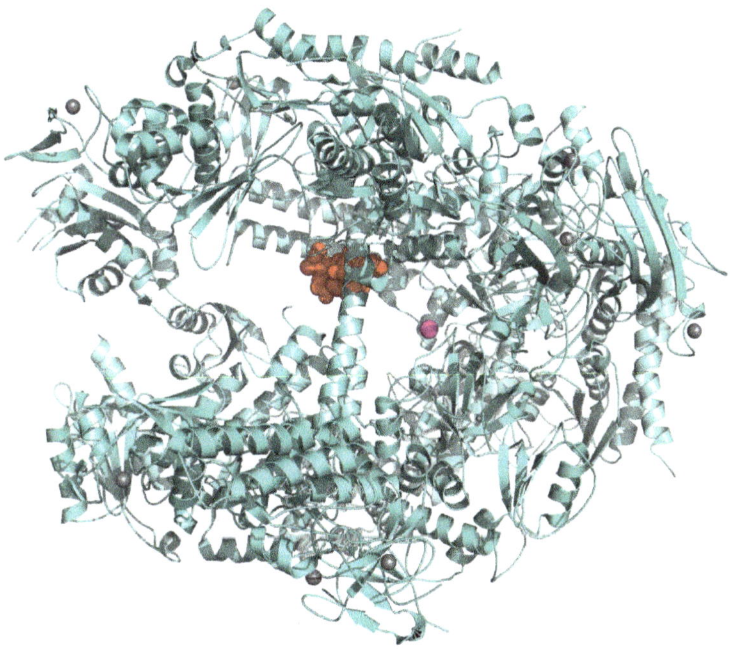

RNA polymerase from *Saccharomyces cerevisiae* complexed with α-amanitin (in red). Despite the use of the term "polymerase," RNA polymerases are classified as a form of nucleotidyl transferase.

A transferase is any one of a class of enzymes that enact the transfer of specific functional groups (e.g. a methyl or glycosyl group) from one molecule (called the donor) to another (called the acceptor). They are involved in hundreds of different biochemical pathways throughout biology, and are integral to some of life's most important processes.

Transferases are involved in myriad reactions in the cell. Three examples of these reactions are the activity of coenzyme A (CoA) transferase, which transfers thiol esters, the action of N-acetyltransferase, which is part of the pathway that metabolizes tryptophan, and the regulation of pyruvate dehydrogenase (PDH), which converts pyruvate to acetyl CoA. Transferases are also utilized during translation. In this case, an amino acid chain is the functional group transferred by a peptidyl transferase. The transfer involves the removal of the growing amino acid chain from the tRNA molecule in the A-site of the ribosome and its subsequent addition to the amino acid attached to the tRNA in the P-site.

Mechanistically, an enzyme that catalyzed the following reaction would be a transferase:

$$Xgroup + Y \xrightarrow[transferase]{} X + Ygroup$$

In the above reaction, X would be the donor, and Y would be the acceptor. "Group" would be the functional group transferred as a result of transferase activity. The donor is often a coenzyme.

Some of the most important discoveries relating to transferases occurred as early as the 1930s. Earliest discoveries of transferase activity occurred in other classifications of enzymes, including Beta-galactosidase, protease, and acid/base phosphatase. Prior to the realization that individual enzymes were capable of such a task, it was believed that two or more enzymes enacted functional group transfers.

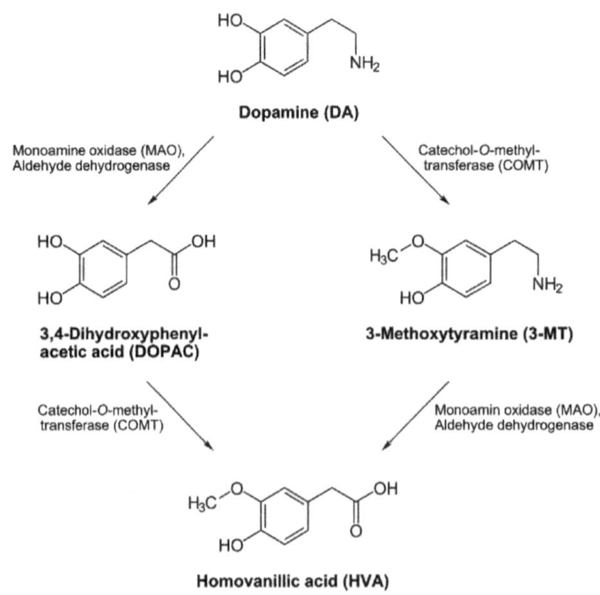

Biodegradation of dopamine via catechol-O-methyltransferase (along with other enzymes). The mechanism for dopamine degradation led to the Nobel Prize in Physiology or Medicine in 1970.

Transamination, or the transfer of an amine (or NH2) group from an amino acid to a keto acid by an aminotransferase (also known as a "transaminase"), was first noted in 1930 by D. M. Needham, after observing the disappearance of glutamic acid added to pigeon breast muscle. This observance was later verified by the discovery of its reaction mechanism by Braunstein and Kritzmann in 1937. Their analysis showed that this reversible reaction could be applied to other tissues.This assertion was validated by Rudolf Schoenheimer's work with radioisotopes as tracers in 1937. This in turn would pave the way for the possibility that similar transfers were a primary means of producing most amino acids via amino transfer.

Another such example of early transferase research and later reclassification involved the discovery of uridyl transferase. In 1953, the enzyme UDP-glucose pyrophosphorylase was shown to be a transferase, when it was found that it could reversibly produce UTP and G1P from UDP-glucose and an organic pyrophosphate.

Another example of historical significance relating to transferase is the discovery of the mechanism of catecholamine breakdown by catechol-O-methyltransferase. This discovery was a large part of the reason for Julius Axelrod's 1970 Nobel Prize in Physiology or Medicine (shared with Sir Bernard Katz and Ulf von Euler).

Classification of transferases continues to this day, with new ones being discovered frequently. An example of this is Pipe, a sulfotransferase involved in the dorsal-ventral patterning of Drosophilia.Initially, the exact mechanism of Pipe was unknown, due to a lack of information on its substrate.Research into Pipe's catalytic activity eliminated the likelihood of it being a heparan sulfate glycosaminoglycan. Further research has shown that Pipe targets the ovarian structures for sulfation. Pipe is currently classified as a Drosophilia heparan sulfate 2-O-sulfotransferase.

Nomenclature

Systematic names of transferases are constructed in the form of "donor:acceptor grouptransferase."For example, methylamine:L-glutamate N-methyltransferase would be the standard naming convention for the transferase methylamine-glutamate N-methyltransferase, where methylamine is the donor, L-glutamate is the acceptor, and methyltransferase is the EC category grouping. This same action by the transferase can be illustrated as follows:

methylamine + L-glutamate $\rightleftharpoons NH_3$ + N-methyl-L-glutamate

However, other accepted names are more frequently used for transferases, and are often formed as "acceptor grouptransferase" or "donor grouptransferase." For example, a DNA methyltransferase is a transferase that catalyzes the transfer of a methyl group to a DNA acceptor. In practice, many molecules are not referred to using this terminology due to more prevalent common names. For example, RNA Polymerase is the modern common name for what was formerly known as RNA nucleotidyltransferase, a kind of nucleotidyl transferase that transfers nucleotides to the 3' end of a growing RNA strand. In the EC system of classification, the accepted name for RNA Polymerase is DNA-directed RNA polymerase.

Classification

Described primarily based on the type of biochemical group transferred, transferases can be divided into ten categories (based on the EC Number classification). These categories comprise over 450 different unique enzymes. In the EC numbering system, transferases have been given a classification of EC2. Hydrogen is not considered a functional group when it comes to transferase targets; instead, hydrogen transfer is included under oxidoreductases, due to electron transfer considerations.

Classification of transferases into subclasses

EC number	Examples	Group(s) transferred
EC 2.1	methyltransferase and formyltransferase	single-carbon groups
EC 2.2	transketolase and transaldolase	aldehyde or ketone groups
EC 2.3	acyltransferase	acyl groups or groups that become alkyl groups during transfer
EC 2.4	glycosyltransferase, hexosyltransferase, and pentosyltransferase	glycosyl groups, as well as hexoses and pentoses
EC 2.5	riboflavin synthase and chlorophyll synthase	alkyl or aryl groups, other than methyl groups
EC 2.6	transaminase, and oximinotransferase	nitrogenous groups
EC 2.7	phosphotransferase, polymerase, and kinase	phosphorus-containing groups; subclasses are based on the acceptor (e.g. alcohol, carboxyl, etc.)
EC 2.8	sulfurtransferase and sulfotransferase	sulfur-containing groups
EC 2.9	selenotransferase	selenium-containing groups
EC 2.10	molybdenumtransferase and tungstentransferase	molybdenum or tungsten

Reactions

EC 2.1: Single Carbon Transferases

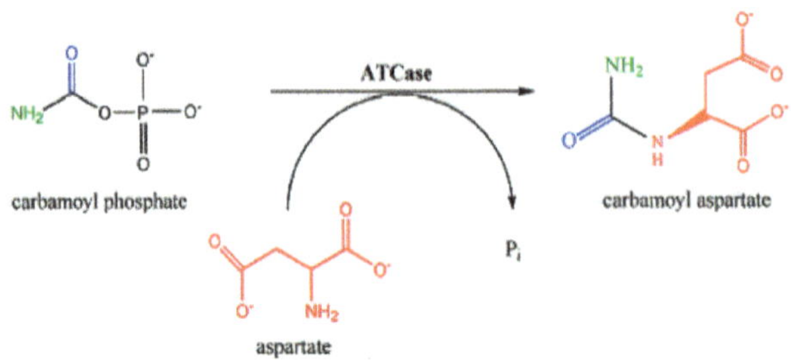

Reaction involving aspartate transcarbamylase.

EC 2.1 includes enzymes that transfer single-carbon groups. This category consists of transfers of methyl, hydroxymethyl, formyl, carboxy, carbamoyl, and amido groups. Carbamoyltransferases, as an example, transfer a carbamoyl group from one molecule to another. Carbamoyl groups follow the formula NH2CO. In ATCase such a transfer is written as Carbamyl phosphate + L-aspertate ⇌ L-carbamyl aspartate + phosphate.

EC 2.2: Aldehyde and Ketone Transferases

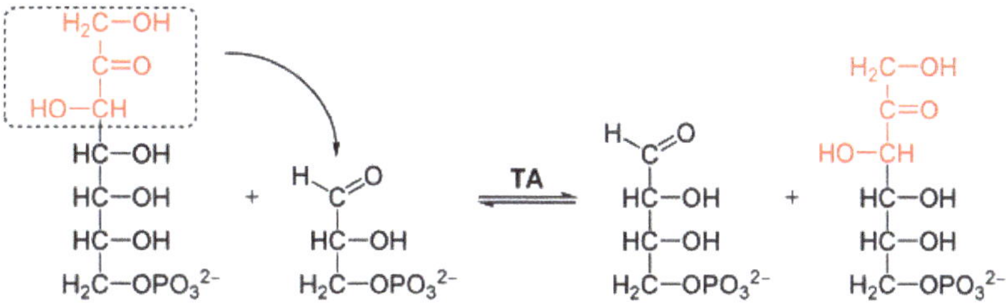

The reaction catalyzed by transaldolase.

Enzymes that transfer aldehyde or ketone groups and included in EC 2.2. This category consists of various transketolases and transaldolases. Transaldolase, the namesake of aldehyde transferases, is an important part of the pentose phosphate pathway. The reaction it catalyzes consists of a transfer of a dihydroxyacetone functional group to Glyceraldehyde 3-phosphate (also known as G3P). The reaction is as follows: sedoheptulose 7-phosphate + glyceraldehyde 3-phosphate ⇌ erythrose 4-phosphate + fructose 6-phosphate.

EC 2.3: acyl Transferases

Transfer of acyl groups or acyl groups that become alkyl groups during the process of being transferred are key aspects of EC 2.3. Further, this category also differentiates between amino-acyl and non-amino-acyl groups. Peptidyl transferase is a ribozyme that facilitates formation of peptide bonds during translation. As an aminoacyltransferase, it catalyzes the transfer of a peptide to an aminoacyl-tRNA, following this reaction: peptidyl-$tRNA_A$ + aminoacyl-$tRNA_B$ ⇌ $tRNA_A$ + peptidyl aminoacyl-$tRNA_B$.

EC 2.4: Glycosyl, Hexosyl, and Pentosyl Transferases

EC 2.4 includes enzymes that transfer glycosyl groups, as well as those that transfer hexose and pentose. Glycosyltransferase is a subcategory of EC 2.4 transferases that is involved in biosynthesis of disaccharides and polysaccharides through transfer of monosaccharides to other molecules. An example of a prominent glycosyltransferase is lactose synthase which is a dimer possessing two protein subunits. Its primary action is to produce lactose from glucose and UDP-galactose. This occurs via the following pathway: UDP-β-D-galactose + D-glucose ⇌ UDP + lactose

EC 2.5: alkyl and aryl transferases

EC 2.5 relates to enzymes that transfer alkyl or aryl groups, but does not include methyl groups. This is in contrast to functional groups that become alkyl groups when transferred, as those are included in EC 2.3. EC 2.5 currently only possesses one sub-class: Alkyl and aryl transferases. Cysteine synthase, for example, catalyzes the formation of acetic acids and cysteine from O3-acetyl-L-serine and hydrogen sulfide: O_3-acetyl-L-serine + $H_2S \rightleftharpoons$ L-cysteine + acetate.

EC 2.6: Nitrogenous Transferases

Aspartate aminotransferase can act on several different amino acids.

The grouping consistent with transfer of nitrogenous groups is EC 2.6. This includes enzymes like transaminase (also known as "aminotransferase"), and a very small number of oximinotransferases and other nitrogen group transferring enzymes. EC 2.6 previously included amidinotransferase but it has since been reclassified as a subcategory of EC 2.1 (single-carbon transferring enzymes). In the case of aspartate transaminase, which can act on tyrosine, phenylalanine, and tryptophan, it reversibly transfers an amino group from one molecule to the other.

The reaction, for example, follows the following order:
L-aspartate +2-oxoglutarate $\rightleftharpoons$ oxaloacetate + L- -glutamate.

EC 2.7: Phosphorus Transferases

While EC 2.7 includes enzymes that transfer phosphorus-containing groups, it also includes nuclotidyl transferases as well. Sub-category phosphotransferase is divided up in categories based on the type of group that accepts the transfer. Groups that are classified as phosphate acceptors include: alcohols, carboxy groups, nitrogenous groups, and phosphate groups. Further constituents of this subclass of transferases are various kinases. A prominent kinase is cyclin-dependent kinase (or CDK), which comprises a sub-family of protein kinases. As their name implies, CDKs are heavily dependent on specific cyclin molecules for activation. Once combined, the CDK-cyclin complex is capable of enacting its function within the cell cycle.

The reaction catalyzed by CDK is as follows: ATP + a target protein {\displaystyle \rightarrow } ADP + a phosphoprotein.

EC 2.8: Sulfur Transferases

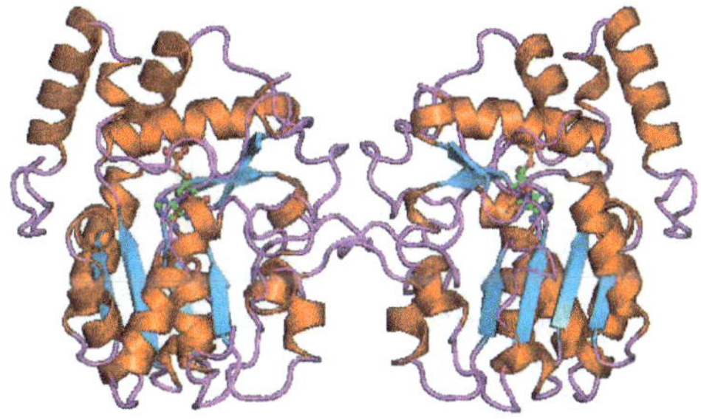

Ribbon diagram of a variant structure of estrogen sulfotransferase (PDB 1aqy EBI)

Transfer of sulfur-containing groups is covered by EC 2.8 and is subdivided into the subcategories of sulfurtransferases, sulfotransferases, and CoA-transferases, as well as enzymes that transfer alkylthio groups. A specific group of sulfotransferases are those that use PAPS as a sulfate group donor. Within this group is alcohol sulfotransferase which has a broad targeting capacity. Due to this, alcohol sulfotransferase is also known by several other names including "hydroxysteroid sulfotransferase," "steroid sulfokinase," and "estrogen sulfotransferase." Decreases in its activity has been linked to human liver disease. This transferase acts via the following reaction: O_3-acetyl-L-serine + $H_2S \rightleftharpoons$ L-cysteine + acetate.

EC 2.9: Selenium Transferases

EC 2.9 includes enzymes that transfer selenium-containing groups. This category only contains two transferases, and thus is one of the smallest categories of transferase. Selenocysteine synthase, which was first added to the classification system in 1999, converts seryl-tRNA(Sec UCA) into selenocysteyl-tRNA(Sec UCA).

EC 2.10: Metal Transferases

The category of EC 2.10 includes enzymes that transfer molybdenum or tungsten-containing groups. However, as of 2011, only one enzyme has been added: molybdopterin molybdotransferase. This enzyme is a component of MoCo biosynthesis in Escherichia coli. The reaction it catalyzes is as follows: adenylyl-molybdopterin + molybdate → molybdenum cofactor + AMP.

Role in Histo-Blood Group

The A and B transferases are the foundation of the human ABO blood group system. Both A and B transferases are glycosyltransferases, meaning they transfer a sugar molecule onto an H-antigen.This allows H-antigen to synthesize the glycoprotein and glycolipid conjugates that are known as the A/B antigens. The full name of A transferase is alpha 1-3-N-acetylgalactosaminyltransferase and its function in the cell is to add N-acetylgalactosamine to H-antigen, creating A-antigen.:55 The full name of B transferase is alpha 1-3-galactosyltransferase, and its function in the cell is to add a galactose molecule to H-antigen, creating B-antigen.

It is possible for Homo sapiens to have any of four different blood types: Type A (express A antigens), Type B (express B antigens), Type AB (express both A and B antigens) and Type O (express neither A nor B antigens). The gene for A and B transferases is located on chromosome nine. The gene contains seven exons and six introns and the gene itself is over 18kb long. The alleles for A and B transferases are extremely similar. The resulting enzymes only differ in 4 amino acid residues. The differing residues are located at positions 176, 235, 266, and 268 in the enzymes.

Deficiencies

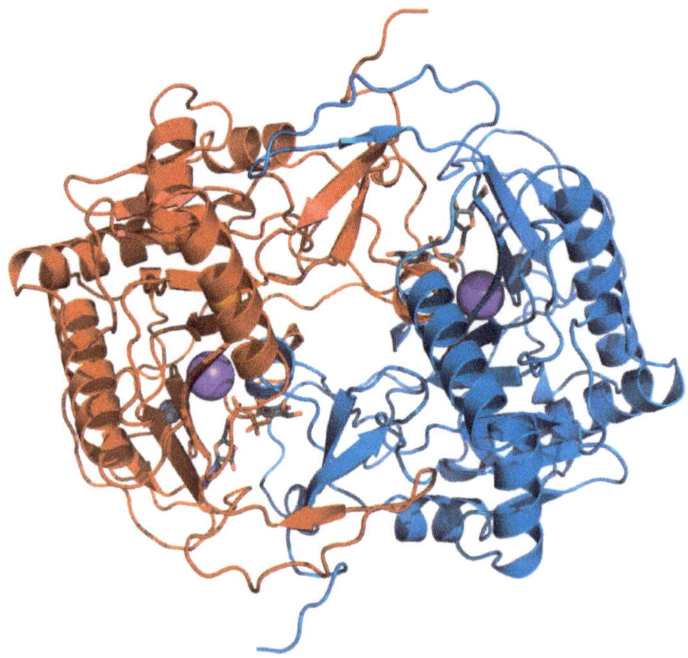

A deficiency of this transferase, E. coli galactose-1-phosphate uridyltransferase is a known cause of galactosemia

Transferase deficiencies are at the root of many common illnesses. The most common result of a transferase deficiency is a buildup of a cellular product.

SCOT Deficiency

Succinyl-CoA:3-ketoacid CoA transferase deficiency (or SCOT deficiency) leads to a buildup of ketones. Ketones are created upon the breakdown of fats in the body and are an important energy source. Inability to utilize ketones leads to intermittent ketoacidosis, which usually first manifests during infancy. Disease sufferers experience nausea, vomiting, inability to feed, and breathing difficulties. In extreme cases, ketoacidosis can lead to coma and death. The deficiency is caused by mutation in the gene OXTC1. Treatments mostly rely on controlling the diet of the patient.

CPT-II Deficiency

Carnitine palmitoyltransferase II deficiency (also known as CPT-II deficiency) leads to an excess long chain fatty acids, as the body lacks the ability to transport fatty acids into the mitochondria to be processed as a fuel source. The disease is caused by a defect in the gene CPT2. This deficiency will present in patients in one of three ways: lethal neonatal, severe infantile hepatocardiomuscular, and myopathic form. The myopathic is the least severe form of the deficiency and can manifest at any point in the lifespan of the patient. The other two forms appear in infancy. Common symptoms of the lethal neonatal form and the severe infantile forms are liver failure, heart problems, seizures and death. The myopathic form is characterized by muscle pain and weakness following vigorous exercise. Treatment generally includes dietary modifications and carnitine supplements.

Galactosemia

Galactosemia results from an inability to process galactose, a simple sugar. This deficiency occurs when the gene for galactose-1-phosphate uridylyltransferase (GALT) has any number of mutations, leading to a deficiency in the amount of GALT produced. There are two forms of Galactosemia: classic and Duarte. Duarte galactosemia is generally less severe than classic galactosemia and is caused by a deficiency of galactokinase. Galactosemia renders infants unable to process the sugars in breast milk, which leads to vomiting and anorexia within days of birth. Most symptoms of the disease are caused by a buildup of galactose-1-phosphate in the body. Common symptoms include liver failure, sepsis, failure to grow, and mental impairment, among others. Buildup of a second toxic substance, galactitol, occurs in the lenses of the eyes, causing cataracts. Currently, the only available treatment is early diagnosis followed by adherence to a diet devoid of lactose, and prescription of antibiotics for infections that may develop.

Choline Acetyltransferase Deficiencies

Choline acetyltransferase (also known as ChAT or CAT) is an important enzyme which produces the neurotransmitter acetylcholine. Acetylcholine is involved in many neuropsychic functions such as memory, attention, sleep and arousal. The enzyme is globular in shape and consists of a single amino acid chain. ChAT functions to transfer an acetyl group from acetyl co-enzyme A to choline in the synapses of nerve cells and exists in two forms: soluble and membrane bound. The ChAT gene is located on chromosome 10.

Alzheimer's Disease

Decreased expression of ChAT is one of the hallmarks of Alzheimer's disease. Patients with Alzheimer's disease show a 30 to 90% reduction in activity in several regions of the brain, including the temporal lobe, the parietal lobe and the frontal lobe. However,

ChAT deficiency is not believed to be the main cause of this disease.

Amyotrophic lateral sclerosis (ALS or Lou Gehrig's disease)

Patients with ALS show a marked decrease in ChAT activity in motor neurons in the spinal cord and brain. Low levels of ChAT activity are an early indication of the disease and are detectable long before motor neurons begin to die. This can even be detected before the patient is symptomatic.

Huntington's Disease

Patients with Huntington's also show a marked decrease in ChAT production. Though the specific cause of the reduced production is not clear, it is believed that the death of medium-sized motor neurons with spiny dendrites leads to the lower levels of ChAT production.

Schizophrenia

Patients with Schizophrenia also exhibit decreased levels of ChAT, localized to the mesopontine tegment of the brain and the nucleus accumbens, which is believed to correlate with the decreased cognitive functioning experienced by these patients.

Sudden Infant Death Syndrome (SIDS)

Recent studies have shown that SIDS infants show decreased levels of ChAT in both the hypothalamus and the striatum. SIDS infants also display fewer neurons capable of producing ChAT in the vagus system. These defects in the medulla could lead to an inability to control essential autonomic functions such as the cardiovascular and respiratory systems.

Congenital Myasthenic Syndrome (CMS)

CMS is a family of diseases that are characterized by defects in neuromuscular transmission which leads to recurrent bouts of apnea (inability to breathe) that can be fatal. ChAT deficiency is implicated in myasthenia syndromes where the transition problem occurs presynaptically. These syndromes are characterized by the patients' inability to resynthesize acetylcholine.

Uses in Biotechnology

Terminal Transferases

Terminal transferases are transferases that can be used to label DNA or to produce plasmid vectors.It accomplishes both of these tasks by adding deoxynucleotides in the form of a template to the downstream end or 3' end of an existing DNA molecule. Terminal transferase is one of the few DNA polymerases that can function without an RNA primer.

Glutathione Transferases

The family of glutathione transferases (GST) is extremely diverse, and therefore can be used for a number of biotechnological purposes. Plants use glutathione transferases as a means to segregate toxic metals from the rest of the cell. These glutathione transferases can be used to create biosensors to detect contaminants such as herbicides and insecticides. Glutathione transferases are also used in transgenic plants to increase resistance to both biotic and abiotic stress. Glutathione transferases are currently being explored as targets for anti-cancer medications due to their role in drug resistance. Further, glutathione transferase genes have been investigated due to their ability to prevent oxidative damage and have shown improved resistance in transgenic cultigens.

Rubber Transferases

Currently the only available commercial source of natural rubber is the Hevea plant (Hevea brasiliensis). Natural rubber is superior to synthetic rubber in a number of commercial uses. Efforts are being made to produce transgenic plants capable of synthesizing natural rubber, including tobacco and sunflower. These efforts are focused on sequencing the subunits of the rubber transferase enzyme complex in order to transfect these genes into other plants.

HYDROLASE

In biochemistry, a hydrolase or hydrolytic enzyme is an enzyme that catalyzes the hydrolysis of a chemical bond. For example, an enzyme that catalyzed the following reaction is a hydrolase:

A–B + H2O → A–OH + B–H

Nomenclature

Systematic names of hydrolases are formed as "substrate hydrolase." However, common names are typically in the form "substratease." For example, a nuclease is a hydrolase that cleaves nucleic acids.

Classification

Hydrolases are classified as EC 3 in the EC number classification of enzymes. Hydrolases can be further classified into several subclasses, based upon the bonds they act upon:

- EC 3.1: ester bonds (esterases: nucleases, phosphodiesterases, lipase, phosphatase)
- EC 3.2: sugars (DNA glycosylases, glycoside hydrolase)
- EC 3.3: ether bonds
- EC 3.4: peptide bonds (Proteases/peptidases)
- EC 3.5: carbon-nitrogen bonds, other than peptide bonds
- EC 3.6 acid anhydrides (acid anhydride hydrolases, including helicases and GTPase)

- EC 3.7 carbon-carbon bonds
- EC 3.8 halide bonds
- EC 3.9: phosphorus-nitrogen bonds
- EC 3.10: sulphur-nitrogen bonds
- EC 3.11: carbon-phosphorus bonds
- EC 3.12: sulfur-sulfur bonds
- EC 3.13: carbon-sulfur bonds

Clinical Considerations

Hydrolase secreted by Lactobacillus jensenii in the human gut stimulates the liver to secrete bile salts that aids in the digestion of food.

Etymology and Pronunciation

The word hydrolase suffixes the combining form of -ase to the hydrol syllables of hydrolysis.

LYASE

In biochemistry, a lyase is an enzyme that catalyzes the breaking (an «elimination" reaction) of various chemical bonds by means other than hydrolysis (a «substitution" reaction) and oxidation, often forming a new double bond or a new ring structure. The reverse reaction is also possible (called a "Michael addition"). For example, an enzyme that catalyzed this reaction would be a lyase:

ATP → cAMP + PPi

Lyases differ from other enzymes in that they require only one substrate for the reaction in one direction, but two substrates for the reverse reaction.

Nomenclature

Systematic names are formed as "substrate group-lyase." Common names include decarboxylase, dehydratase, aldolase, etc. When the product is more important, synthase may be used in the name, e.g. phosphosulfolactate synthase (EC 4.4.1.19, Michael addition of sulfite to phosphoenolpyruvate). A combination of both an elimination and a Michael addition is seen in O-succinylhomoserine (thiol)-lyase (MetY or MetZ) which catalyses first the γ-elimination of O-succinylhomoserine (with succinate as a leaving group) and then the addition of sulfide to the vinyl intermediate, this reaction was first classified as a lyase (EC 4.2.99.9), but was then reclassified as a transferase (EC 2.5.1.48).

Classification

Lyases are classified as EC 4 in the EC number classification of enzymes. Lyases can be further classified into seven subclasses:

- EC 4.1 includes lyases that cleave carbon-carbon bonds, such as decarboxylases (EC 4.1.1), aldehyde lyases (EC 4.1.2), oxo acid lyases (EC 4.1.3), and others (EC 4.1.99)
- EC 4.2 includes lyases that cleave carbon-oxygen bonds, such as dehydratases
- EC 4.3 includes lyases that cleave carbon-nitrogen bonds
- EC 4.4 includes lyases that cleave carbon-sulfur bonds
- EC 4.5 includes lyases that cleave carbon-halide bonds
- EC 4.6 includes lyases that cleave phosphorus-oxygen bonds, such as adenylyl cyclase and guanylyl cyclase
- EC 4.99 includes other lyases, such as ferrochelatase

ISOMERASE

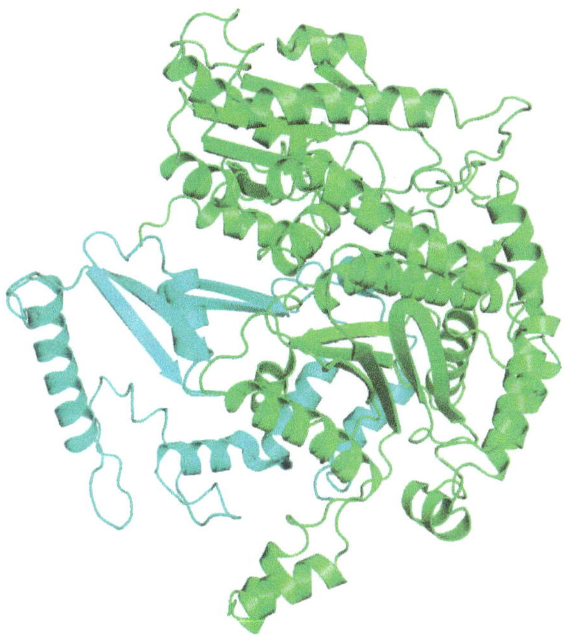

Ribbon diagram of Glucose-6-Phosphate isomerase.

Isomerases are a general class of enzymes which convert a molecule from one isomer to another. Isomerases can either facilitate intramolecular rearrangements in which bonds are broken and formed. The general form of such a reaction is as follows:

A–B → B–A

There is only one substrate yielding one product. This product has the same molecular formula as the substrate but differs in bond connectivity or spatial arrangements. Isomerases catalyze reactions across many biological processes, such as in glycolysis and carbohydrate metabolism.

Isomerization

Examples of Isomers

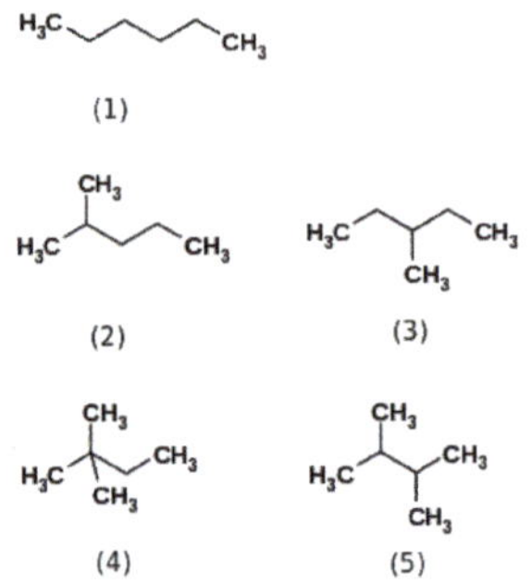

The structural isomers of hexane.

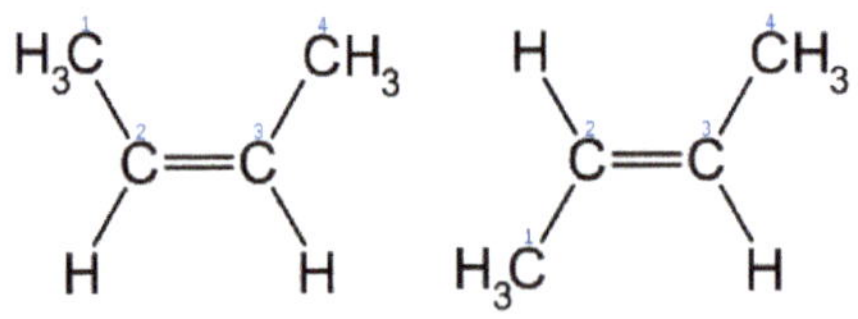

Cis-2-butene and Trans-2-butene.

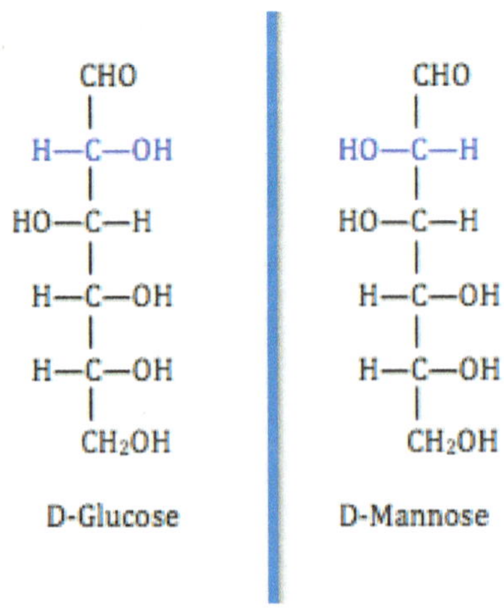

Epimers: D-glucose and D-mannose.

Isomerases catalyze changes within one molecule. They convert one isomer to another, meaning that the end product has the same molecular formula but a different physical structure. Isomers themselves exist in many varieties but can generally be classified as structural isomers or stereoisomers. Structural isomers have a different ordering of bonds and/or different bond connectivity from one another, as in the case of hexane and its four other isomeric forms (2-methylpentane, 3-methylpentane, 2,2-dimethylbutane, and 2,3-dimethylbutane).

Stereoisomers have the same ordering of individual bonds and the same connectivity but the three-dimensional arrangement of bonded atoms differ. For example, 2-butene exists in two isomeric forms: cis-2-butene and trans-2-butene. The sub-categories of isomerases containing racemases, epimerases and cis-trans isomers are examples of enzymes catalyzing the interconversion of stereoisomers. Intramolecular lyases, oxidoreductases and transferases catalyze the interconversion of structural isomers.

The prevalence of each isomer in nature depends in part on the isomerization energy, the difference in energy between isomers. Isomers close in energy can interconvert easily and are often seen in comparable proportions. The isomerization energy, for example, for converting from a stable cis isomer to the less stable trans isomer is greater than for the reverse reaction, explaining why in the absence of isomerases or an outside energy source such as ultraviolet radiation a given cis isomer tends to be present in greater amounts than the trans isomer. Isomerases can increase the reaction rate by lowering the isomerization energy.

Calculating isomerase kinetics from experimental data can be more difficult than for other enzymes because the use of product inhibition experiments is impractical. That is, isomerization is not an irreversible reaction since a reaction vessel will contain one substrate and one product so the typical simplified model for calculating reaction kinetics does not hold. There are also practical difficulties in determining the rate-determining step at high concentrations in a single isomerization. Instead, tracer perturbation can overcome these technical difficulties if there are two forms of the unbound enzyme. This technique uses isotope exchange to measure indirectly the interconversion of the free enzyme between its two forms. The radiolabeled substrate and product diffuse in a time-dependent manner. When the system reaches equilibrium the addition of unlabeled substrate perturbs or unbalances it. As equilibrium is established again, the radiolabeled substrate and product are tracked to determine energetic information.

The earliest use of this technique elucidated the kinetics and mechanism underlying the action of phosphoglucomutase, favoring the model of indirect transfer of phosphate with one intermediate and the direct transfer of glucose. This technique was then adopted to study the profile of proline racemase and its two states: the form which isomerizes L-proline and the other for D-proline. At high concentrations it was shown that the transition state in this interconversion is rate-limiting and that these enzyme forms may differ just in the protonation at the acidic and basic groups of the active site.

Nomenclature

Generally, "the names of isomerases are formed as "substrate isomerase" (for example, enoyl CoA isomerase), or as "substrate type of isomerase" (for example, phosphoglucomutase)."

Classification

Enzyme-catalyzed reactions each have a uniquely assigned classification number. Isomerase-catalyzed reactions have their own EC category: EC 5. Isomerases are further classified into six subclasses:

Racemases, Epimerases

This category (EC 5.1) includes (racemases) and epimerases). These isomerases invert stereochemistry at the target chiral carbon. Racemases act upon molecules with one chiral carbon for inversion of stereochemistry, whereas epimerases target molecules with multiple chiral carbons and act upon one of them. A molecule with only one chiral carbon has two enantiomeric forms, such as serine having the isoforms D-serine and L-serine differing only in the absolute configuration about the chiral carbon. A molecule with multiple chiral carbons has two forms at each chiral carbon. Isomerization at one chiral carbon of several yields epimers, which differ from one another in absolute configuration at just one chiral carbon. For example, D-glucose and D-mannose differ in configuration at just one chiral carbon. This class is further broken down by the group the enzyme acts upon:

Racemases and epimerases:		
EC number	Description	Examples
EC 5.1.1	Acting on Amino Acids and Derivative	alanine racemase, methionine racemase
EC 5.1.2	Acting on Hydroxy Acids and Derivatives	lactate racemase, tartrate epimerase
EC 5.1.3	Acting on Carbohydrates and Derivatives	ribulose-phosphate 3-epimerase, UDP-glucose 4-epimerase
EC 5.1.99	Acting on Other Compounds	methylmalonyl CoA epimerase, hydantoin racemase

Cis-trans Isomerases

This category (EC 5.2) includes enzymes that catalyze the isomerization of cis-trans isomers. Alkenes and cycloalkanes may have cis-trans stereoisomers. These isomers are not distinguished by absolute configuration but rather by the position of substituent groups relative to a plane of reference, as across a double bond or relative to a ring structure. Cis isomers have substituent groups on the same side and trans isomers have groups on opposite sides.

This category is not broken down any further. All entries presently include:

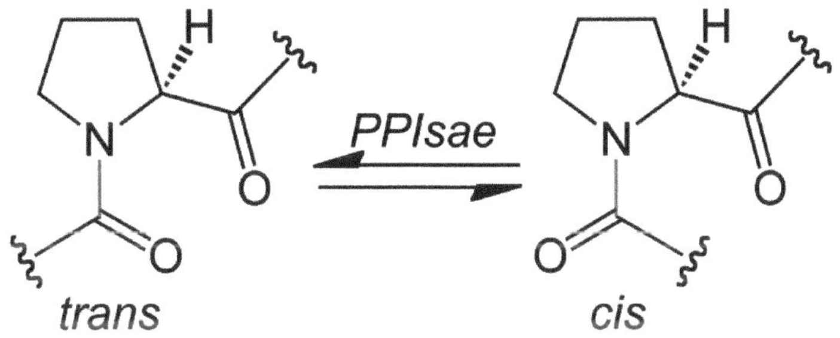

conversion mediated by peptidylprolyl isomerase.

Cis-trans isomerases:	
EC number	**Examples**
EC 5.2.1.1	Maleate isomerase
EC 5.2.1.2	Maleylacetoacetate isomerase
EC 5.2.1.4	Maleylpyruvate isomerase
EC 5.2.1.5	Linoleate isomerase
EC 5.2.1.6	Furylfuramide isomerase
EC 5.2.1.8	Peptidylprolyl isomerase
EC 5.2.1.9	Farnesol 2-isomerase
EC 5.2.1.10	2-chloro-4-carboxymethylenebut-2-en-1,4-olide isomerase
EC 5.2.1.12	Zeta-carotene isomerase
EC 5.2.1.13	Prolycopene isomerase
EC 5.2.1.14	Beta-carotene isomerase

Intramolecular Oxidoreductases

This category (EC 5.3) includes intramolecular oxidoreductases. These isomerases catalyze the transfer of electrons from one part of the molecule to another. In other words, they catalyze the oxidation of one part of the molecule and the concurrent reduction of another part. Sub-categories of this class are:

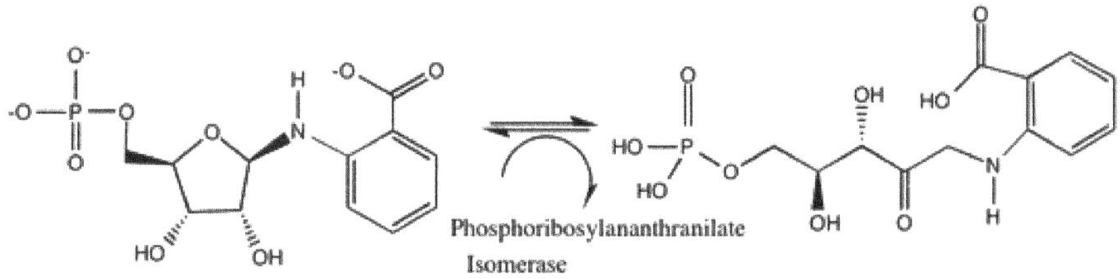

reaction catalyzed by phosphoribosylanthranilate isomerase.

Intramolecular oxidoreductases:		
EC number	Description	Examples
EC 5.3.1	Interconverting Aldoses and Ketoses	Triose-phosphate isomerase, Ribose-5-phosphate isomerase
EC 5.3.2	Interconverting Keto- and Enol- Groups	Phenylpyruvate tautomerase, Oxaloacetate tautomerase
EC 5.3.3	Transposing C=C Double Bonds	Steroid Delta-isomerase, L-dopachrome isomerase
EC 5.3.4	Transposing S-S Bonds	Protein disulfide-isomerase
EC 5.3.99	Other Intramolecular Oxidoreductases	Prostaglandin-D synthase, Allene-oxide cyclase

Intramolecular Transferases

This category (EC 5.4) includes intramolecular transferases (mutases). These isomerases catalyze the transfer of functional groups from one part of a molecule to another. Phosphotransferases (EC 5.4.2) were categorized as transferases (EC 2.7.5) with regeneration of donors until 1983. This sub-class can be broken down according to the functional group the enzyme transfers:

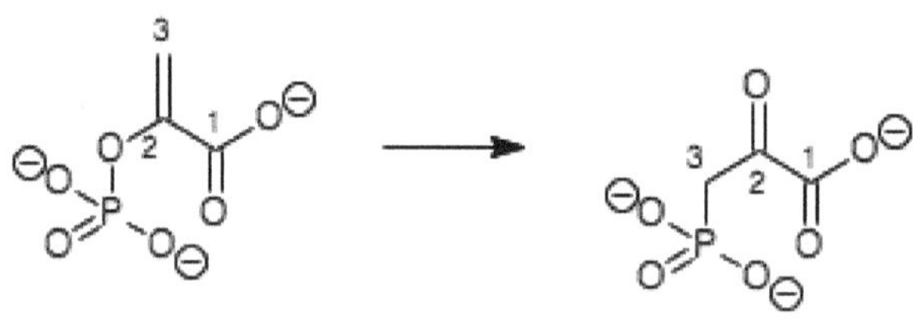

reaction catalyzed by phosphoenolpyruvate mutase.

Intramolecular transferases:		
EC number	Description	Examples
EC 5.4.1	Transferring Acyl Groups	Lysolecithin acylmutase, Precorrin-8X methylmutase
EC 5.4.2	Phosphotransferases (Phosphomutases)	Phosphoglucomutase, Phosphopentomutase
EC 5.4.3	Transferring Amino Groups	Beta-lysine 5,6-aminomutase, Tyrosine 2,3-aminomutase
EC 5.4.4	Transferring hydroxy groups	(hydroxyamino)benzene mutase, Isochorismate synthase

EC 5.4.99	Transferring Other Groups	Methylaspartate mutase, Chorismate mutase

Intramolecular Lyases

This category (EC 5.5) includes intramolecular lyases. These enzymes catalyze "reactions in which a group can be regarded as eliminated from one part of a molecule, leaving a double bond, while remaining covalently attached to the molecule." Some of these catalyzed reactions involve the breaking of a ring structure.

This category is not broken down any further. All entries presently include:

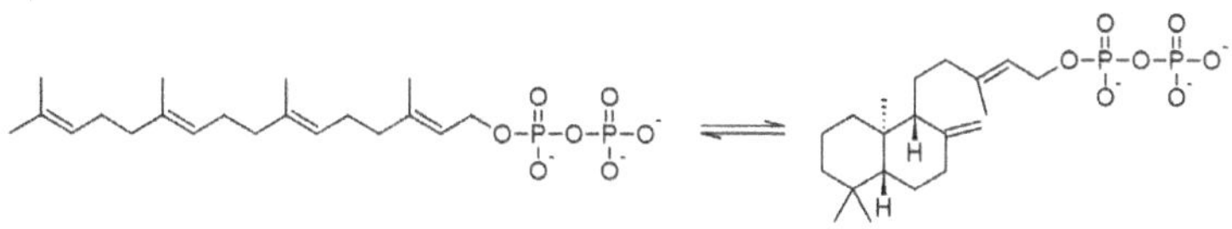

reaction catalyzed by ent-Copalyl diphosphate synthase

Intramolecular lyases:

EC number	Examples
EC 5.5.1.1	Muconate cycloisomerase
EC 5.5.1.2	3-carboxy-cis,cis-muconate cycloisomerase
EC 5.5.1.3	Tetrahydroxypteridine cycloisomerase
EC 5.5.1.4	Inositol-3-phosphate synthase
EC 5.5.1.5	Carboxy-cis,cis-muconate cyclase
EC 5.5.1.6	Chalcone isomerase
EC 5.5.1.7	Chloromuconate cycloisomerase
EC 5.5.1.8	(+)-bornyl diphosphate synthase
EC 5.5.1.9	Cycloeucalenol cycloisomerase
EC 5.5.1.10	Alpha-pinene-oxide decyclase
EC 5.5.1.11	Dichloromuconate cycloisomerase
EC 5.5.1.12	Copalyl diphosphate synthase
EC 5.5.1.13	Ent-copalyl diphosphate synthase
EC 5.5.1.14	Syn-copalyl-diphosphate synthase
EC 5.5.1.15	Terpentedienyl-diphosphate synthase

EC 5.5.1.16	Halimadienyl-diphosphate synthase
EC 5.5.1.17	(S)-beta-macrocarpene synthase
EC 5.5.1.18	Lycopene epsilon-cyclase
EC 5.5.1.19	Lycopene beta-cyclase
EC 5.5.1.20	Prosolanapyrone-III cycloisomerase
EC 5.5.1.n1	D-ribose pyranase

Mechanisms of Isomerases

Ring expansion and Contraction Via Tautomers

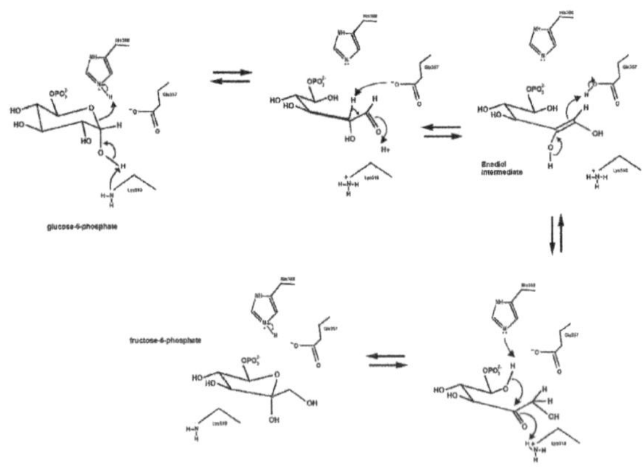

The isomerization of glucose-6-phosphate by glucose-6-phosphate isomerase.

A classic example of ring opening and contraction is the isomerization of glucose (an aldehyde with a six-membered ring) to fructose (a ketone with a five-membered ring). The conversion of D-glucose-6-phosphate to D-fructose-6-phosphate is catalyzed by glucose-6-phosphate isomerase, an intramolecular oxidoreductase. The overall reaction involves the opening of the ring to form an aldose via acid/base catalysis and the subsequent formation of a cis-endiol intermediate. A ketose is then formed and the ring is closed again.

Glucose-6-phosphate first binds to the active site of the isomerase. The isomerase opens the ring: its His388 residue protonates the oxygen on the glucose ring (and thereby breaking the O5-C1 bond) in conjunction with Lys518 deprotonating the C1 hydroxyl oxygen. The ring opens to form a straight-chain aldose with an acidic C2 proton. The C3-C4 bond rotates and Glu357 (assisted by His388) depronates C2 to form a double bond between C1 and C2. A cis-endiol intermediate is created and the C1 oxygen is protonated by the catalytic residue, accompanied by the deprotonation of the endiol C2 oxygen. The straight-chain ketose is formed. To close the fructose ring, the reverse of ring opening occurs and the ketose is protonated.

Epimerization

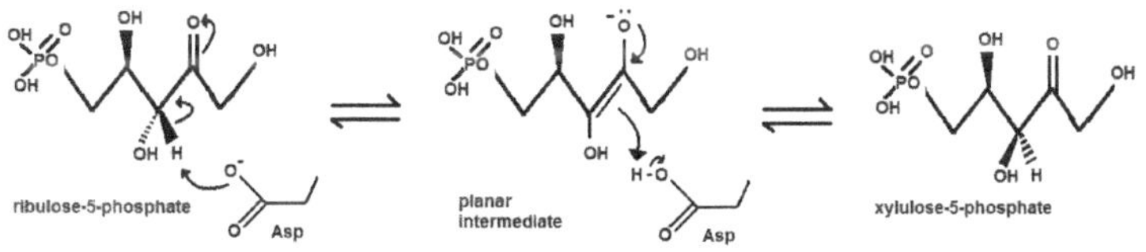

The conversion of ribulose-5-phosphate to xylulose-5-phosphate.

An example of epimerization is found in the Calvin cycle when D-ribulose-5-phosphate is converted into D-xylulose-5-phosphate by ribulose-phosphate 3-epimerase. The substrate and product differ only in stereochemistry at the third carbon in the chain. The underlying mechanism involves the deprotonation of that third carbon to form a reactive enolate intermediate. The enzyme's active site contains two Asp residues. After the substrate binds to the enzyme, the first Asp deprotonates the third carbon from one side of the molecule. This leaves a planar sp2-hybridized intermediate. The second Asp is located on the opposite side of the active side and it protonates the molecule, effectively adding a proton from the back side. These coupled steps invert stereochemistry at the third carbon.

Intramolecular Transfer

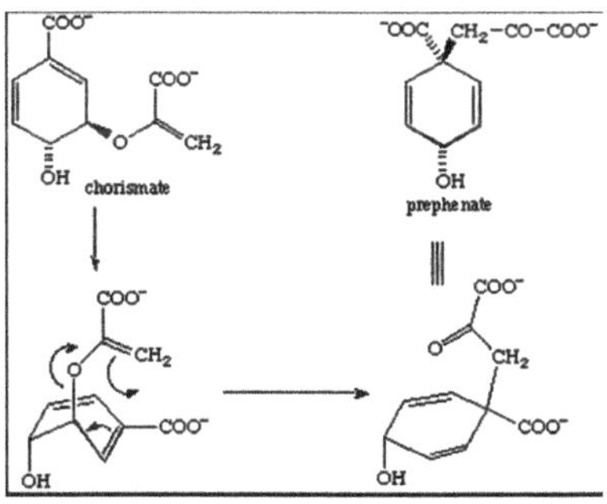

A proposed mechanism for chorismate mutase. Clark, T., Stewart, J.D. and Ganem, B. Transition-state analogue inhibitors of chlorismate mutase. Tetrahedron 46 (1990) 731-748. © IUBMB 2001.

Chorismate mutase is an intramolecular transferase and it catalyzes the conversion of chorismate to prephenate, used as a precursor for L-tyrosine and L-phenylalanine in some plants and bacteria. This reaction is a Claisen rearrangement that can proceed with or without the isomerase, though the rate increases 106 fold in the presence of chorismate mutase. The reaction goes through a chair transition state with the substrate in a trans-diaxial position. Experimental evidence indicates that the isomerase selectively binds the chair transition state, though the exact mechanism of catalysis is not known. It is thought that this binding stabilizes the transition state through electrostatic effects, accounting for the dramatic increase in the reaction rate in the presence of the mutase or upon addition of a specifically-placed cation in the active site.

Intramolecular Oxidoreduction

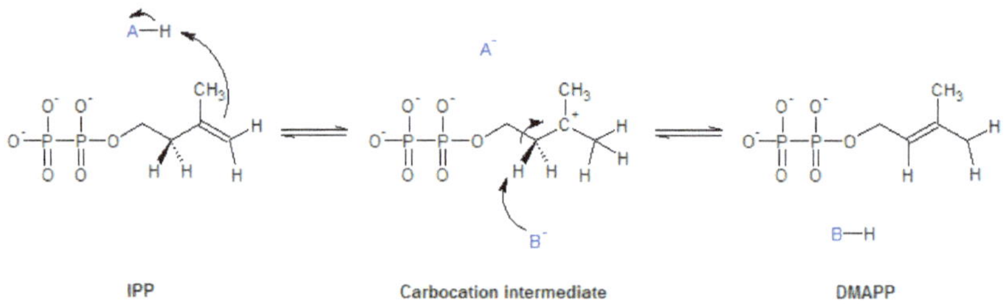

Conversion by IPP isomerase.

Isopentenyl-diphosphate delta isomerase type I (also known as IPP isomerase) is seen in cholesterol synthesis and in particular it catalyzes the conversion of isopentenyl diphosphate (IPP) to dimethylallyl diphosphate (DMAPP). In this isomerization reaction a stable carbon-carbon double bond is rearranged top create a highly electrophilic allylic isomer. IPP isomerase catalyzes this reaction by the stereoselective antarafacial transposition of a single proton. The double bond is protonated at C4 to form a tertiary carbocation intermediate at C3. The adjacent carbon, C2, is deprotonated from the opposite face to yield a double bond. In effect, the double bond is shifted over.

The role of isomerase in human disease

Isomerase plays a role in human disease. Deficiencies of this enzyme can cause disorders in humans.

Phosphohexose Isomerase Deficiency

Phosphohexose Isomerase Dificiency (PHI) is also known as phosphoglucose isomerase deficiency or Glucose-6-phosphate isomerase deficiency, and is a hereditary enzyme deficiency. PHI is the second most frequent erthoenzyopathy in glycolysis besides pyruvate kinase deficiency, and is associated with non-spherocytic haemolytic anaemia of variable severity. This disease is centered on the glucose-6-phosphate protein. This protein can be found in the secretion of some cancer cells. PHI is the result of a dimeric enzyme that catalyses the reversible interconversion of fructose-6-phosphate and gluose-6-phosphate.

PHI is a very rare disease with only 50 cases reported in literature to date.

Diagnosis is made on the basis of the clinical picture in association with biochemical studies revealing erythrocyte GPI deficiency (between 7 and 60% of normal) and identification of a mutation in the GPI gene by molecular analysis.

The deficiency of phosphohexose isomerase can lead to a condition referred to as hemolytic syndrome. As in humans, the hemolytic syndrome, which is characterized by a diminished erythrocyte number, lower hematocrit, lower hemoglobin, higher number of reticulocytes and plasma bilirubin concentration, as well as increased liver- and spleen-somatic indices, was exclusively manifested in homozygous mutants.

Triosephosphate Isomerase Deficiency

The disease referred to as triosephosphate isomerase deficiency (TPI), is a severe autosomal recessive inherited multisystem disorder of glycolyic metabolism. It is characterized by hemolytic anemia and neurodegeneration, and is caused by anaerobic metabolic dysfunction. This dysfunction results from a missense mutation that effects the encoded TPI protein. The most common mutation is the substitution of gene, Glu104Asp, which produces the most severe phenotype, and is responsible for approximately 80% of clinical TPI deficiency.

TPI deficiency is very rare with less than 50 cases reported in literature. Being an autosomal recessive inherited disease, TPI deficiency has a 25% recurrence risk in the case of heterozygous parents. It is a congenital disease that most often occurs with hemolytic anemia and manifests with jaundice. Most patients with TPI for Glu104Asp mutation or heterozygous for a TPI null allele and Glu104Asp have a life expectancy of infancy to early childhood. TPI patients with other mutations generally show longer life expectancy. To date, there are only two cases of individuals with TPI living beyond the age of 6. These cases involve two brothers from Hungary, one who did not develop neurological symptoms until the age of 12, and the older brother who has no neurological symptoms and suffers from anemia only.

Individuals with TPI show obvious symptoms after 6–24 months of age. These symptoms include: dystonia, tremor, dyskinesia, pyramidal tract signs, cardiomyopathy and spinal motor neuron involvement. Patients also show frequent respiratory system bacterial infections.

TPI is detected through deficiency of enzymatic activity and the build-up of dihyroxyacetone phosphate(DHAP), which is a toxic substrate, in erythrocytes. This can be detected through physical examination and a series of lab work. In detection, there is generally myopathic changes seen in muscles and chronic axonal neuropathy found in the nerves. Diagnosis of TPI can be confirmed through molecular genetics. Chorionic villus DNA analysis or analysis of fetal red cells can be used to detect TPI in antenatal diagnosis.

Treatment for TPI is not specific, but varies according to different cases. Because of the range of symptoms TPI causes, a team of specialist may be needed to provide treatment to a single individual. That team of specialists would consists of pediatricians, cardiologists, neurologists, and other healthcare professionals, that can develop a comprehensive plan of action.

Supportive measures such as red cell transfusions in cases of severe anaemia can be taken to treat TPI as well. In some cases, spleen removal (splenectomy) may improve the anaemia. There is no treatment to prevent progressive neurological impairment of any other non-haematological clinical manifestation of the diseases.

Industrial Applications

By far the most common use of isomerases in industrial applications is in sugar manufacturing. Glucose isomerase (also known as xylose isomerase) catalyzes the conversion of D-xylose and D-glucose to D-xylulose and D-fructose. Like most sugar isomerases, glucose isomerase catalyzes the interconversion of aldoses and ketoses.

The conversion of glucose to fructose is a key component of high-fructose corn syrup production. Isomerization is more specific than older chemical methods of fructose production, resulting in a higher yield of fructose and no side products. The fructose produced from this isomerization reaction is purer with no residual flavors from contaminants. High-fructose corn syrup is preferred by many confectionery and soda manufacturers because of the high sweetening power of fructose (twice that of sucrose), its relatively low cost and its inability to crystallize. Fructose is also used as a sweetener for use by diabetics. Major issues of the use of glucose isomerase involve its inactivation at higher temperatures and the requirement for a high pH (between 7.0 and 9.0) in the reaction environment. Moderately high temperatures, above 70 °C, increase the yield of fructose by at least half in the isomerization step.The enzyme requires a divalent cation such as Co2+ and Mg2+ for peak activity, an additional cost to manufacturers. Glucose isomerase also has a much higher affinity for xylose than for glucose, necessitating a carefully controlled environment.

The isomerization of xylose to xylulose has its own commercial applications as interest in biofuels has increased. This reaction is often seen naturally in bacteria that feed on decaying plant matter. Its most common industrial use is in the production of ethanol, achieved by the fermentation of xylulose. The use of hemicellulose as source material is very common. Hemicellulose contains xylan, which itself is composed of xylose in β(1,4) linkages. The use of glucose isomerase very efficiently converts xylose to xylulose, which can then be acted upon by fermenting yeast. Overall, extensive research in genetic engineering has been invested into optimizing glucose isomerase and facilitating its recovery from industrial applications for re-use.

Glucose isomerase is able to catalyze the isomerization of a range of other sugars, including D-ribose, D-allose and L-arabinose. The most efficient substrates are those similar to glucose and xylose, having equatorial hydroxyl groups at the third and fourth carbons. The current model for the mechanism of glucose isomerase is that of a hydride shift based on X-ray crystallography and isotope exchange studies.

LIGASE

In biochemistry, a ligase is an enzyme that can catalyze the joining of two large molecules by forming a new chemical bond, usually with accompanying hydrolysis of a small pendant chemical group on one of the larger molecules or the enzyme catalyzing the linking together of two compounds, e.g., enzymes that catalyze joining of C-O, C-S, C-N, etc. In general, a ligase catalyzes the following reaction:

Ab + C → A–C + b

or sometimes

Ab + cD → A–D + b + c + d + e + f

where the lowercase letters signify the small, dependent groups. Ligase can join two complementary fragments of nucleic acid and repair single stranded breaks that arise in double stranded DNA during replication.

Nomenclature

The common names of ligases often include the word "ligase", such as DNA ligase, an enzyme commonly used in molecular biology laboratories to join together DNA fragments. Other common names for ligases include the word "synthetase", because they are used to synthesize new molecules.

Biochemical nomenclature has sometimes distinguished synthetases from synthases and sometimes treated the words as synonymous. Under one definition, synthases do not use energy from nucleoside triphosphates (such as ATP, GTP, CTP, TTP, and UTP), whereas synthetases do use nucleoside triphosphates. It is also said that a synthase is a lyase (a lyase is an enzyme that catalyzes the breaking of various chemical bonds by means other than hydrolysis and oxidation, often forming a new double bond or a new ring structure) and does not require any energy, whereas a synthetase is a ligase (a ligase is an enzyme that binds two chemicals or compounds) and thus requires energy. However, the Joint Commission on Biochemical Nomenclature (JCBN) dictates that "synthase" can be used with any enzyme that catalyses synthesis (whether or not it uses nucleoside triphosphates), whereas "synthetase" is to be used synonymously.

Classification

Ligases are classified as EC 6 in the EC number classification of enzymes. Ligases can be further classified into six subclasses:

- EC 6.1 includes ligases used to form carbon-oxygen bonds
- EC 6.2 includes ligases used to form carbon-sulfur bonds
- EC 6.3 includes ligases used to form carbon-nitrogen bonds (including argininosuccinate synthetase)
- EC 6.4 includes ligases used to form carbon-carbon bonds
- EC 6.5 includes ligases used to form phosphoric ester bonds
- EC 6.6 includes ligases used to form nitrogen-metal bonds

Etymology and Pronunciation

The word ligase uses combining forms of lig- (from the Latin verb ligāre, "to bind" or "to tie together") + -ase (denoting an enzyme), yielding "binding enzyme".

REFERENCES

1. Enzyme nomenclature, 1978 recommendations of the Nomenclature Committee of the International Union of Biochemistry on the nomenclature and classification of enzymes. New York: Academic Press. 1979. ISBN 9780323144605.
2. McNaught AD (1997). Compendium of Chemical Terminology (2nd ed.). Oxford: Blackwell Scientific Publications. ISBN 0-9678550-9-8.
3. Whitesell JK, Fox MA (2004). Organic Chemistry (3rd ed.). Sudbury, Mass.: Jones and Bartlett. pp. 220–222. ISBN 978-0-7637-2197-8.
4. Cornish-Bowden A. Fundamentals of Enzyme Kinetics (4th ed.). Weinheim: Wiley-VCH. pp. 238–241. ISBN 978-3-527-66548-8.
5. Fisher LM, Albery WJ, Knowles JR (May 1986). "Energetics of proline racemase: tracer perturbation experiments using [14C]proline that measure the interconversion rate of the two forms of free enzyme". Biochemistry. 25 *(9): 2538–42.* doi:10.1021/bi00357a038. PMID 3521737.
6. Britton HG, Clarke JB (Nov 1968). "The mechanism of the phosphoglucomutase reaction. Studies on rabbit muscle phosphoglucomutase with flux techniques". The Biochemical Journal. 110 *(2): 161–80.* doi:10.1042/bj1100161. PMC 1187194. PMID 5726186.
7. Bruice PY (2010). Essential Organic Chemistry (2nd ed.). Upper Saddle River, N.J.: Prentice Hall. ISBN 978-0-321-59695-6.
8. Webb EC (1992). Enzyme nomenclature 1992 : recommendations of the Nomenclature Committee of the International Union of Biochemistry and Molecular Biology on the nomenclature and classification of enzymes (6th ed.). San Diego: Published for the International Union of Biochemistry and Molecular Biology by Academic Press. ISBN 978-0-12-227164-9.
9. The Enzyme List Class 5 - Isomerases (PDF). Nomenclature Committee of the International Union of Biochemistry and Molecular Biology (NC-IUBMB). 2010.
10. Solomons JT, Zimmerly EM, Burns S, Krishnamurthy N, Swan MK, Krings S, Muirhead H, Chirgwin J, Davies C (Sep 2004). "The crystal structure of mouse phosphoglucose isomerase at 1.6A resolution and its complex with glucose 6-phosphate reveals the catalytic mechanism of sugar ring opening". Journal of Molecular Biology. 342 *(3): 847–60. doi:10.1016/j.jmb.2004.07.085. PMID 15342241.*
11. Terada T, Mukae H, Ohashi K, Hosomi S, Mizoguchi T, Uehara K (Apr 1985). "Characterization of an enzyme which catalyzes isomerization and epimerization of D-erythrose 4-phosphate". European Journal of Biochemistry / FEBS. 148 *(2): 345–51. doi:10.1111/j.1432-1033.1985.tb08845.x. PMID 3987693.*
12. Bugg T (2012). "Chapter 10: Isomerases". Introduction to Enzyme and Coenzyme Chemistry (3rd ed.). Wiley. ISBN 978-1-118-34896-3.
13. Kast P, Grisostomi C, Chen IA, Li S, Krengel U, Xue Y, Hilvert D (Nov 2000). "A strategically positioned cation is crucial for efficient catalysis by chorismate

mutase". The Journal of Biological Chemistry. 275 *(47): 36832–8. doi:10.1074/jbc.M006351200. PMID 10960481.*

14. Zheng W, Sun F, Bartlam M, Li X, Li R, Rao Z (Mar 2007). "The crystal structure of human isopentenyl diphosphate isomerase at 1.7 A resolution reveals its catalytic mechanism in isoprenoid biosynthesis". Journal of Molecular Biology. 366 *(5): 1447–58. doi:10.1016/j.jmb.2006.12.055. PMID 17250851.*
15. Kugler W, Lakomek M (Mar 2000). "Glucose-6-phosphate isomerase deficiency". Baillière's Best Practice & Research. Clinical Haematology. 13 *(1): 89–101. doi:10.1053/beha.1999.0059. PMID 10916680.*
16. Merkle S, Pretsch W (1993). "Glucose-6-phosphate isomerase deficiency associated with nonspherocytic hemolytic anemia in the mouse: an animal model for the human disease" (PDF). Blood. 81 *(1): 206–13. PMID 8417789.*
17. Krone W, Schneider G, Schulz D, Arnold H, Blume KG (1 January 1970). "Detection of phosphohexose isomerase deficiency in human fibroblast cultures". Humangenetik. 10 *(3): 224–30.* doi:10.1007/BF00295784. PMID 5475507.
18. Orosz PF. "Triose phosphate-isomerase deficiency". Orphanet. Retrieved 14 November 2013.
19. Celotto AM, Frank AC, Seigle JL, Palladino MJ (Nov 2006). "Drosophila model of human inherited triosephosphate isomerase deficiency glycolytic enzymopathy". Genetics. 174 *(3): 1237–46. doi:10.1534/genetics.106.063206. PMC 1667072. PMID 16980388.*
20. Oláh J, Orosz F, Keserü GM, Kovári Z, Kovács J, Hollán S, Ovádi J (Apr 2002). "Triosephosphate isomerase deficiency: a neurodegenerative misfolding disease" (PDF). Biochemical Society Transactions. 30 *(2): 30–8.* doi:10.1042/bst0300030. PMID 12023819.
21. Hollán S, Fujii H, Hirono A, Hirono K, Karro H, Miwa S, Harsányi V, Gyódi E, Inselt-Kovács M (Nov 1993). "Hereditary triosephosphate isomerase (TPI) deficiency: two severely affected brothers one with and one without neurological symptoms". Human Genetics. 92 *(5): 486–90.* doi:10.1007/bf00216456. PMID 8244340.
22. «Triosephosphate Isomerase Deficiency". NORD. Retrieved 14 December 2013.
23. «Triose phosphate isomerase deficiency -TPI" (PDF). Retrieved 26 November 2013.
24. Bhosale SH, Rao MB, Deshpande VV (Jun 1996). "Molecular and industrial aspects of glucose isomerase". Microbiological Reviews. 60 *(2): 280–300. PMC 239444. PMID 8801434.*
25. Baker S (1976). "Pure fructose syrups". Process Biochemistry. 11: 20–25.
26. Antrim RL, Colilla W, Schnyder BJ (1979). "Glucose isomerase production of high fructose syrups". Applied Biochemistry and Bioengineering. 2: 97–155.
27. Wang PY, Shopsis C, Schneider H (May 1980). "Fermentation of a pentose by yeasts". Biochemical and Biophysical Research Communications. 94 *(1): 248–54.* doi:10.1016/s0006-291x(80)80213-0. PMID 6446306.

28. Chen WP (August–September 1980). "Glucose isomerase". Process Biochemistry. 15: 36–41.

Index

S

T